# Duelling Idiots
### and
### Other Probability
# Puzzlers

PAUL J. NAHIN

Duelling Idiots
and Other Probability
Puzzlers

PRINCETON UNIVERSITY PRESS
PRINCETON AND OXFORD

**Library of Congress Cataloging-in-Publication Data**
Nahin, Paul J.
Duelling idiots and other probability puzzlers / Paul J. Nahin.
p.    cm.
Includes index.
ISBN 0-691-00979-1 (cloth : alk. paper)
1. Probabilities—Problems, exercises, etc.   I. Title.
QA273.N29   2000
519.2—dc21   99-087353

This book has been composed in Goudy and Marydale (display)
Text Design by Carmina Alvarez

The paper used in this publication meets the minimum requirements of
ANSI/NISO Z39.48-1992 (R1997) (Permanence of Paper)

http://pup.princeton.edu

Printed in the United States of America

1   3   5   7   9   10   8   6   4   2

*Frontis*    The author on May 26, 1964, at age twenty-three, sitting in front of his first completed digital design that wasn't a homework assignment (and it actually worked). The huge cabinet was purchased by the U.S. Air Force for use as the programmable telemetry signal simulator of the Gemini space capsule. (Gemini was the second phase of the three-phase Mercury/Gemini/Apollo American space program.) Using J-K/NAND logic, running at a clock speed of one-quarter megahertz, and about 12,000 transistors (a single modern Pentium chip contains millions of transistors), the simulator had a one-half kilobyte magnetic-core memory that took two people to lift. Modern operating systems like Windows were only a fantasy in Bill Gates's preadolescent dreams, and the gadget was programmed in machine language. It had less computing power than most electronic toys have today. Installed at the Eastern Test Range in Cape Canaveral, Florida, the simulator was built as part of a larger project by the now-defunct Systems Division of the Beckman Instruments Company in Fullerton, California, using the probabilistic management tool called PERT, which is discussed in the final problem of this book.

To Patricia Ann, who, thirty-seven years ago, played an "idiots' duel" with me when we married (because, at age 22, who could possibly know what a chance they are taking?). I know I won that game, but she tells me she did, too—which is why I love her.

"I think you're begging the question," said Haydock, "and I can see looming ahead one of those terrible exercises in probability where six men have white hats and six men have black hats and you have to work it out by mathematics how likely it is that the hats will get mixed up and in what proportion. If you start thinking about things like that, you would go round the bend. Let me assure you of that!"

—from *The Mirror Crack'd*, by Agatha Christie (1962)

"The true logic of this world is the calculus of probabilities."

—James Clerk Maxwell (1831–1879), theoretical physicist extraordinaire, in a paraphrase of Pierre Simon de Laplace's (1749–1827) famous assertion in the introduction to his *Théorie Analytique des Probabilitiés* (various editions, 1812–1825).

"Fate, Time, Occasion, Chance and Change? To these
All things are subject . . . "

—from *Prometheus Unbound* (1820), Act II, Scene 4 (lines 119–120), by Percy Bysshe Shelley (1792–1822)

# Contents

# Acknowledgments

There are a number of individuals who greatly aided my writing of this book: Nan Collins at the University of New Hampshire, who typed it; Trevor Lipscombe, my editor at Princeton University Press who sent me a contract; my students at the Universities of Virginia and New Hampshire, who worked through many of the problems as homework and exam questions; my wonderful cats Heaviside and Maxwell, who sat, slept, and purred on the working manuscript (and so kept me company) while I was writing it; David Rutledge at Caltech and John Molinder at Harvey Mudd College, who read the book in typescript and gave me the benefit of their comments; Deborah Wenger, who copy-edited the manuscript; Bob Brown at Princeton, who guided the book through final production; and my wife Patricia, who had to listen to me bellow in frustration every time I misplaced a MATLAB file on the hard drive because I am a clumsy Windows user. I thank them all.

# Preface

In 1965, after spending nearly three years as a designer of digital machines in the Systems Division of Beckman Instruments, Inc., in Fullerton, California, my employment suddenly ceased. The sad reason for this was euphemistically called by upper management a "change of business opportunities." That is, the digital machinery product line was immediately terminated for a lack of any new customers and I was, at the age of twenty-five, looking for a new job.

Fortunately for me, Hughes Aircraft Company had its Ground Systems Group (the division that made ground- and ship-based radars) in the same town, and its management was hiring young electrical engineers. So in December 1965, I became a member of the technical staff at Hughes—where I quickly learned that Boolean algebra and sequential switching theory, which had held me in good stead as a digital systems designer at Beckman, simply wasn't going to be enough for long-term survival as a radar systems analyst at Hughes. I needed to learn some more math. I needed to learn probability theory, and I needed to learn it fast.

As astonishing as it is now to look back on those days, in 1962 I had graduated from Stanford University—one of the world's great schools—with a B.S. degree in electrical engineering without having taken a single course in probability theory. It wasn't that I was lazy, as no one in my class of electrical engineers (EEs) took such a course. It simply wasn't required, and my advisor had never brought it up, even as a suggestion, because everybody thought of probability theory as graduate-level course work. And when, in 1963, I graduated with my master's degree from Caltech, which most people consider to be a veritable hothouse of techno-nerds, it had been neither required nor suggested that a first-year graduate student in electrical engineering study probability.

It was only when I started my doctoral studies in electrical engineering at the Irvine campus of the University of California as a Howard Hughes Staff Doctoral Fellow in 1968 that I took a formal probability course in a degree program. But by then I had been at Hughes for nearly three years and had already started such studies myself, for the most practical of reasons: in order to keep my job.

Actually, even while still at Beckman, I had been exposed to a famous probability question, although I hadn't recognized it as such at the time. It was a twice-a-day routine for groups of the engineering staff to take what was jokingly called a "roach-coach" break. That is, several of us (let's say $N$ people) would, morning and afternoon, take ten minutes to walk out to the parking lot and buy donuts and coffee from the visiting lunch van. Rather than each of us individually paying for our own purchases, however, we played a game called "odd-man-out": each of us would simultaneously flip a coin and then show all the others what we had gotten. If it turned out that everybody but one had the same result ($N - 1$ heads and one tail, or $N - 1$ tails and one head) then the "odd" person paid for everyone. Otherwise we all flipped again, and so on until we got an odd man out.

There are several interesting questions about this game,

but one of immediate practical interest concerns how long it will take, as a function of $N$, to get an odd man out. That is, how many flipping attempts (on average) will it take to reach a decision on who pays? And what if one of the coins is biased? You will see in Problem 9 that it is actually quite easy to calculate the answers once we have established some fundamental theoretical results.

I played "odd-man-out" from 1963 to 1965, all through my stay at Beckman (Hughes was a much more formal place, and I never saw anybody play for donuts and coffee during my six years there), but it never even occurred to me that such questions had answers. And if it had I wouldn't have known how to find them. Today, of course, such an admission would be considered tragic. I presently teach sophomore electrical engineering students at the University of New Hampshire the same material that is in this book, which I didn't see until years after getting a Caltech master's degree. So educational times have changed for the better.

Even easier than calculating the answer, however, is using a computer to simulate the physical situation. By the time you have finished this book, you'll have seen a number of examples of just how to do probabilistic simulation. Indeed, a major, two-part thesis of mine that forms the central pedagogical theme of the book is the following:

PART I: *No matter how smart you are, there will always be a problem harder than one you can solve analytically.*

PART II: *If you know how to use a computer application[1] like MATLAB,[2] you may still be able to solve that "too-hard" problem by simulation.*

I arrived at this thesis the hard way, by direct experience, described in Problem 17. That problem had its origin at

Hughes in terms of an aircraft flying through a region of space that is being watched by radar. In Problem 17 I've stripped away all the technical gingerbread and reduced my description to the mathematical essentials. The physical details of the problem are unimportant here; what matters is that I was assigned to think about that probabilistic problem and given just forty-eight hours to come up with an answer. Unfortunately, in 1968 I had not the faintest idea how to determine it. One thing that I did know, however, was that I definitely was not going to learn how to do it in just forty-eight hours. But, no matter, I still managed to come up with an answer that satisfied my boss (and perhaps kept me employed). How did I do that?

I simulated the problem on a computer, using the simple time-share BASIC system available to all the staff at Hughes. (I can still remember the oily smell of the clanking mechanical connections in the ASR-33 teletypes that served as terminals. How do I love the smoothly operating, oil-odor-free, quiet keyboard of my modern laptop? Let an endless MATLAB loop count the ways!) That is, I wrote a BASIC program that used a uniform random number generator (see the essay at the end of the book for a description of these) to produce numbers at random. Then, with those numbers, the program used geometrical equations I had written into the code to model the physics of the problem. The program then looped around and around and simulated the process 100,000 times. (This took a while in 1968.) From those 100,000 (random) simulations, it was then easy to answer the probability question that was my assigned task.

That's how I saved my job, or at least my pride—by reducing a mathematically impossible problem (for me, in 1968) to a so-called Monte Carlo simulation (in honor, of course, of the famous casino in Monaco). I had, quite literally, used a computer and related software to perform as "digital dice" to get my numerical results. That is what a large part of this book is all about: how to expand your problem-solving power not only by using the theory of probability, but also by seeing how to create computer simulations to use when your personal theoretical problem-solving engine starts to run out of gas.

The concept of using probabilistic simulation to solve physical problems is generally credited to the brilliant Polish mathematician Stanislaw Ulam (1909–1984). Ulam was a key player in the American atomic bomb project (code-named the Manhattan District Project) during the war years 1943 to 1945 at Los Alamos, New Mexico. After the war, he continued to make seminal contributions to the development of the American hydrogen fusion bomb (the H-bomb). The A-bomb project was an effort that required the solution of an enormous number of extraordinarily difficult problems, many of which had an intense mathematical nature. In his autobiography,[3] Ulam describes how the idea of random simulation came to him:

> The idea for what was later called the Monte Carlo method occurred to me when I was playing solitaire during [an] illness. I noticed that it may be much more practical to get an idea of the probability of the successful outcome of a solitaire game . . . by laying down the cards, or experimenting with the process and merely noticing what proportion comes out successfully, rather than to try to compute all the combinatorial possibilities. . . . This is intellectually surprising, and if not exactly humiliating, it gives one a feeling of modesty about the limits of rational or traditional thinking.

So, from a sick man playing a lonely game of cards came one of the great ideas of computational physics.[4]

As Ulam went on to write,

> In a sufficiently complicated problem, actual sampling is better than an examination of all the chains of possibilities. It occurred to me then that this could be equally true of all processes involving branching of events, as in the production and further multiplication of neutrons in some kind of material containing uranium or other fissile elements [i.e., Ulam is talking of the now famous "chain reaction" concept]. At each stage of the process, there are many possibilities determining the fate of the neutron. It can scatter at one angle, change its velocity, be absorbed, or produce more neutrons by fission of the target nucleus, and so on. The elementary probabilities for each of these possibilities are individually known. . . . But the problem is to know what a succession and branching of perhaps hundreds or thousands or millions will do. . . . The [Monte Carlo] idea was to try out thousands of such possibilities and, at each stage, to select by chance, by means of a "random number" with suitable probability, the fate [of a neutron]. After examining the possible histories of only a few thousand, one will have a good sample and an approximate answer to the problem.

Ulam then wrote what were to prove to be prophetic words: "All one needed was to have the means of producing such sample histories. It so happened that computing machines were coming into existence, and here was something suitable for machine calculation."

Some writers make a curious academic distinction: For

them, "Monte Carlo" refers to the use of random numbers to solve an inherently nonprobabilistic problem (see Problem 5 for two examples), while the probabilistic simulation of a probabilistic problem (such as my Hughes problem) is "just" a simulation. I will not make such a distinction here; for me, it's the use of random numbers that makes a problem "Monte Carlo," and nothing else matters. As partial justification for my position, I will point out that Ulam's own original inspiration for Monte Carlo, a probabilistic game of cards, would not qualify as a Monte Carlo problem under such an academic distinction.

## Notes and References

1. This is not a book on programming. I assume from the start that you are familiar with some high-level computer application, such as BASIC, FORTRAN, PASCAL, C++, or whatever. All these tools use, as does MATLAB, English-like commands that are almost self-explanatory—e.g., *while*, *if*, and *for* loops. (BASIC's *go to* statement, however, is generally thought by computer science cognoscenti to be one of the world's great evils, and you'll see no such thing in MATLAB.) In general, I'm not going to make a big deal out of the nuances of MATLAB unless I do something special. Then I'll tell you some more about MATLAB. Moreover, programming is a bit of an art form and nearly everybody has his or her own special way of doing things. I'm pretty casual about such matters; you do it your way, I'll do it mine, and what counts in the end is that we both get the same (correct) answers. For me, the biggest thing is to include lots of comments so you can figure out, after two weeks of not looking at it, what your code is doing. To write a compact, super-tight program that compresses a hundred hours of thought into nine lines of code, and runs faster than a bat out of hell, used to be an exercise in bravado in the old days when one-half-megahertz clocks were fast and 16K memories were big. But in today's world, where

home computers are closing in on the Holy Grail clock speed of one gigahertz, vast, multiple-megabyte RAMs are commonplace, and disk memory is metaphorically cheaper than dirt, why not write code with lots of chatty commentary that even an idiot can read? It doesn't cost you anything and, after all, we can all be idiots from time to time when reading our own old programs. I speak from experience.

2. MATLAB (for MATrix LABoratory) is an extremely powerful and diverse collection of numerical and symbolic manipulation programs. Its fundamental entity is the matrix—even scalars are thought of as being $1 \times 1$ matrices—which allows some very nifty code to be written. I will, in fact, use very little of that power in this book, but I will take advantage of MATLAB's impressive plotting capabilities to create all the plots in the book. Still, to be honest, MATLAB is an interpretative system (it is not compiled into machine language, executable code prior to running on your machine) and it is possible to achieve great reductions in run times by using MATLAB's matrix nature. I refer all readers with an interest in such matters to specialized works, such as the nice tutorial appendix "Programming in MATLAB" in J.H. McClellan, R. W. Schafer and M. A. Yoder, *DSP First: A Multimedia Approach* (New York: Prentice Hall, 1998).

3. S. M. Ulam, *Adventures of a Mathematician* (Charles Scribner's Sons, 1976), pp. 196–98. See also *From Cardinals to Chaos: Reflections on the Life and Legacy of Stanislaw Ulam* (New York: Cambridge University Press, 1989).

4. An historical purist might object to naming Ulam as the inventor of Monte Carlo, as one can find the anecdotal use of random numbers in solving physical problems in much earlier literature. In particular, the great Scottish engineer/scientist William Thomson, later Lord Kelvin (1824–1907), used the drawing of numbers at random (written on slips of paper and pulled out of a hat) in a paper published before 1900. But it was Ulam, fifty years later, who started the scientific study of the Monte Carlo idea itself to the point that it is today a recognized specialty of computational physics.

# Duelling Idiots
### and
### Other Probability
# Puzzlers

# Introduction

## Part 1

This is a book for people who really like probability problems. There are, I think, a lot of people who fall into that category. Indeed, the editors of *Parade*, a magazine insert in millions of Sunday newspapers across America, thought a probabilistic question intriguing enough to put it on the cover of their issue of August 10, 1997. For the real connoisseur of probability, however, it was actually a pretty tame problem: "Your dog has a litter of four. Is it most likely that two are males and two are females?"

That question was posed in the "Ask Marilyn" column by the famously intelligent Marilyn vos Savant, who answered, "Nope! The most likely split is three males and one female, or three females and one male." That is correct, too, for the case of female/male births being equally likely. Two males and two females has a probability of $\frac{3}{8}$, while the second case has probability of $\frac{1}{2}$. Vos Savant, who has carved a successful writing career partly out of posing old math questions (with answers that have been known for centuries) to readers who find them new, doesn't give her fans the math behind the answer, but for this book, the doggy problem is just too elementary to be included as a legitimate "probability puzzler."

Please don't misunderstand me. While I occasionally think Vos Savant is just a bit too unrevealing of her debt to ancient mathematical lore (I suspect that many of her readers think she is the originator of the problems in her column), I do think she does provide a useful service by publishing such problems. Who could deny that it is a refreshing change to see math of any sort in a newspaper column, as compared to the more typical, seemingly endless rehashing of the supposed details of celebrity lives, or other similar sophomoric speculations? Vos Savant's column is written in the spirit of Laplace's famous dictum, "The theory of probabilities is at bottom nothing but common sense reduced to calculus," but, while clever, Laplace did overstate his argument just a bit. Once beyond the doggy type of question, probability theory can quickly become nonintuitive in the extreme, even for experienced analysts. That is, perhaps surprisingly, one of its most seductive and charming features.

It isn't hard to understand vos Savant's reluctance to put real math in her column, of course, as the quote from Agatha Christie at the beginning of this book accurately reflects how most math-innocents view technical analyses of any sort. As W. Somerset Maugham wrote in the first paragraph of his short story "Mr. Harrington's Washing," "Man has always found it easier to sacrifice his life than to learn the multiplication table." But if you have read this far, then you certainly don't fall into that category. The doggy problem is so simple (see Part II of this introduction) that one could literally write down all sixteen possibilities (sixteen, because there are four consecutive births, each with two possible outcomes, and $2 \times 2 \times 2 \times 2 = 16$) and then just count how many times each of the different situations occurs. In the jargon of mathematics, "we have several different events defined on a finite sample space, with each sample point the result of a Bernoulli sequence of four trials, with the probability of a success being one-half."

Now that is a mouthful. And I am not going to define any of those terms; if you know what that last sentence means, then you just passed the test qualifying you to get the most out of this book. If you don't know what that sentence said, then this book may be just a bit too much for you, at least for now. But that doesn't mean you shouldn't buy it. Do buy it and use it as a study supplement as you take an elementary course in probability.

As I declared before, the doggy problem is really just a routine drill problem, the sort of question that textbook authors put a dozen or so of at the end of each section of their books. Such problems are important to do as learning exercises, and every beginning student should do a number of them when first learning any new math topic. There are lots of such problems in all of the mostly excellent probability textbooks available today; so many, in fact, that Vos Savant will never run out of recyclable drill problems with which to dazzle her readers.

Once beyond the drill problem stage, however, most probability students are eager to try their new and powerful skills on more challenging, more interesting problems. That is the sort of problem you'll find in this book. And where did these problems come from, you may wonder. During the past twenty-five years, I have taught (and continue to teach) probability theory to undergraduate electrical engineering students at the University of New Hampshire. (My debt to hundreds of students who have patiently listened to me talk and scribble on the blackboard in EE647 is a very large one, indeed.) At the end of each term there are always pleas to provide some sort of extra credit work with which to bolster grades, and I have responded by offering what I call "Challenge Problems." These are optional problems (students have to accept the challenge before seeing the problem, and after seeing it, they can't change their minds) to be done as "take-homes" during the week before the final exam (independent work only), with unlimited time, and no partial credit. If a student gets the problem right, then I add five points to his or her final exam score. But if he or she gets it wrong, then I subtract five points.

Over the last twenty-five years, I have created perhaps a hundred or so such questions, and the ones I think are the best are included here. None, to my knowledge, has ever appeared in print before, at least not in the way posed here. The level of these problems is elementary, but that simply means that they all can be done with no mathematics beyond freshman calculus (and at least one of them can be solved with just arithmetic). Each problem has a detailed solution and extended discussion (often including computer illustrations using the powerful scientific application software called MATLAB) in the second half of the book. The problems are much like the famous Birthday Problem or the Buffon Needle Problem, neither of which is included here because they have become so easy to find in textbooks. The problems here are actually no more conceptually difficult than are those two classics, however, and I hope you have fun trying your luck on them.

## Part II.

### Binary Numbers, the Doggy Problem, the Gulf War, and Shooting at Targets

The doggy problem can be solved by simply counting. If we let 1 denote a female birth and 0 denote a male birth, then the sixteen possibilities for a litter of four dogs are represented by the sixteen four-digit binary numbers from 0 to 15:

| | |
|---|---|
| 0 = 0000 | 8 = 1000 |
| 1 = 0001 | 9 = 1001 |
| 2 = 0010 | 10 = 1010 |
| 3 = 0011 | 11 = 1011 |
| 4 = 0100 | 12 = 1100 |
| 5 = 0101 | 13 = 1101 |
| 6 = 0110 | 14 = 1110 |
| 7 = 0111 | 15 = 1111. |

As a mathematician would put it, "Each number is a sample point in the sample space of the experiment of having a litter of four puppies." If a male and female are equally likely to be born, then each sample point is equally likely, with a probability of $\frac{1}{16}$. There are six sample points with two 1's and two 0's (the numbers 3, 5, 6, 9, 10, and 12), and there are eight sample points with either three 1's and one 0 or one 1 and three 0's (the numbers 7, 11, 13, and 14, and the numbers 1, 2, 4, and 8, respectively). Thus, the probability of two males and two females is $\frac{6}{16} = \frac{3}{8}$, and the probability of either three males and one female or one male and three females is $\frac{8}{16} = \frac{1}{2}$.

It was surprising to me that as elementary as this counting technique is, it at first appeared as if Marilyn vos Savant had been unaware of it. I say this because in her *Parade* column of June 14, 1998, she printed a letter asking her a question (the details of which are not important here); the correspondent ended by informing vos Savant that he was also asking "my other two heroes—Stephen Hawking and Kurt Vonnegut—[the same question]. I figure that my chances are 7 out of 8 that I will get at least one response from the three of you." To that, vos Savant replied, "I have a question for you: How in the world did you figure those chances?!"

The answer to Marilyn's question is an even simpler binary counting exercise than is the doggy problem. Just let 1 denote a response and 0 denote no response, and we see immediately that there are eight possibilities; the binary numbers 000 (no responses at all) to 111 (three responses). Seven of those eight binary numbers have at least one 1, so if the correspondent was assuming each of the eight sample points is equally likely (the probability of each response is $\frac{1}{2}$ and that the responses are independent of each other), we then have the stated chances by inspection. If the probability of a response is not $\frac{1}{2}$, then the chances are different but no more difficult to calculate. Suppose $p$ is the probability of a response from a hero; thus, $1 - p$ is the probability of no response. Then the probability of no responses at all is $(1 - p)^3$ and thus the probability of at least one response is $1 - (1 - p)^3$. For example, if $p = \frac{1}{6}$, then the chances for at least one response are 91 out of 216.

My original impression that vos Savant was unaware of all this turned out to be wrong, however, because in her September 6, 1998, column in *Parade*, she addressed the problem again by printing a letter from a reader asking just how the original correspondent calculated his odds. Marilyn's answer made it clear that she hadn't really meant to imply that she didn't know how to count in binary, as she correctly listed all eight possible combinations of reply possibilities from herself, Hawking, and Vonnegut. But then she stumbled again, with the following words about the original correspondent:

> Then he incorrectly figured that these eight possi-
> bilities were equally likely. This is a lot like saying

there are two possibilities regarding sunrise tomorrow: (1) The sun will rise in the morning; or (2) the sun will not rise in the morning. And, therefore, the chances are only fifty-fifty that the sun will rise!

Marilyn's first sentence is simply not a valid objection, and the rest of that passage is just irrelevant. The original correspondent certainly did not "figure incorrectly" by assuming each reply had a probability of $\frac{1}{2}$. The probability of each reply could be anything from 0 to 1, and he was well within his rights to make the special (if perhaps overly optimistic) assumption of $\frac{1}{2}$. Vos Savant's remarks concerning the sun are beside the point: The probabilities of the individual replies are due to individual human decisions, while the event of the sun rising tomorrow is the result of the physical laws of gravity and orbital mechanics. Probability has nothing to do with it. But, curiously, her particular imagery is reminiscent of Laplace's famously incorrect use of Bayes's theorem of conditional probability to compute the odds of the sun's rising tomorrow. Could this be the case of yet another classic puzzler that she once read about, and of which she has since forgotten the proper historical setting? (Laplace presented this calculation in his famous 1814 "Essai Philosophique des Probabilités.")

The counting method works well for problems that involve a small number of different possibilities, but in general we need a more powerful approach. The sophisticated math behind the doggy problem is simply the binomial theorem applied to a Bernoulli sequence of trials (which is a sequence with two characteristics: The trials are independent, and each trial has precisely two possible outcomes). If there are $n$ births, with the probability of a female birth being $p$ (and so the probability of a male birth is $1 - p$), then the probability of $k$ females and $n - k$ males is given by

$$\binom{n}{k} p^k (1 - p)^{n - k}$$

where the binomial coefficient $\binom{n}{k} = \frac{n!}{(n - k)! k!}$ denotes the number of different ways of selecting $k$ things from $n$ things. The factorial function is defined for positive integers as $n! = n(n - 1)(n - 2) \ldots (2)(1)$. Notice, too, that since $\binom{n}{n} = 1$, i.e., there is just one way to select all $n$ things, then the binomial coefficient formula reduces to the special and important result that $0! = 1$, *not* the zero that beginning students so often write.

For the doggy problem we have $n = 4$ and $p = \frac{1}{2}$. So, the probability of two females ($k = 2$) and two males is

$$\binom{4}{2} \left(\frac{1}{2}\right)^2 \left(\frac{1}{2}\right)^2 = \frac{4!}{2! \, 2!} \cdot \frac{1}{2^4} = \frac{6}{16} = \frac{3}{8},$$

and the probability of either one female ($k = 1$) and three males, or three females ($k = 3$) and one male, is

$$\binom{4}{1} \left(\frac{1}{2}\right) \left(\frac{1}{2}\right)^3 + \binom{4}{3} \left(\frac{1}{2}\right)^3 \left(\frac{1}{2}\right) = \frac{4!}{1! 3!} \cdot \frac{1}{16} + \frac{4!}{3! 1!} \cdot$$

$$\frac{1}{16} = \frac{4}{16} + \frac{4}{16} = \frac{1}{2}.$$

A far more interesting application of this simple math was reported in the *Boston Globe* on January 24, 1992 (p. 3), as part of the retrospective analyses then being conducted on the Gulf War. In particular, the Pentagon had gone on record with a claim that the Patriot antiaircraft missile system had "successfully engaged over 80 percent" of the Scud missiles Iraq had launched at Saudi Arabia. An MIT physicist, Theodore Postol, disputed that claim; it was a remarkable claim, too, as the Patriot was designed to counter relatively slow manned aircraft, not supersonic ballistic missiles.

Postol based his skepticism on what he saw after watching videotapes of fourteen Patriot-Scud engagements. There were thirteen misses and one probable hit. The *Globe* article ended with this quote from Professor Postol: "What are the odds I would see 13 misses and one hit if the Patriot was successfully shooting down 80 percent of the Scuds?"

No answer was given in the newspaper, but we can easily calculate it for ourselves using the doggy-problem math. Simply think of a hit as having the claimed probability of 0.8 (thus, a miss has a probability of 0.2), and if we assume independent engagements, we thus have

$$\binom{14}{1} (0.8) (0.2)^{13}$$

as the probability that Professor Postol would see what he saw. The numbers work out to give a probability of less than $10^{-8}$, a value so small that most people would reject the Pentagon's claim of $p = 0.8$. Flipping a fair coin and getting twenty-six consecutive heads is more likely. That is, if the Pentagon's claim had merit, then an event of incredibly low probability had been observed to actually occur: an event with odds of more than a hundred million to one against it. Even most megabuck lotteries have better odds than that. It is, in fact, far more likely that the Pentagon claim did not have merit.

(An aside: I have come to call such rare occurrences *Daphne events*, after Daphne Tams, the contracts and copyright manager for Princeton University Press, for whom it once snowed on seventeen (!) consecutive birthdays. This would be most incredible, of course, if her birthday were in the middle of summer, but even for a winter birthday, it seems to be an a priori rare event. For more on Daphne's snowy birthdays, see Problem 5.)

I have used Postol's problem as lecture material on Bernoulli trials simply because it is "interesting" (and I'll leave that word undefined and just say students seem to find the military context alone fascinating). Calculating the answer to Postol's question involves no technical difficulties, and it can be done on the most elementary of scientific hand calculators; I am using the one I keep in my office desk drawer for emergencies, one I bought for fifteen dollars a couple of years ago in a discount drugstore. However, another aspect of Bernoulli problems that I emphasize in class is that they can easily evolve into questions that, while easy to set up, can present calculation challenges. Don't forget that the goal for engineers, scientists, and applied mathematicians is the calculation of an answer. If you can't do the final calculation you've failed, no matter how pretty your preliminary analysis may be.

As my final example, then, consider the following problem. You are told that a markswoman can hit a target with each shot from her rifle with a fixed probability $p$ that is, before she starts shooting, equally likely to be $\frac{1}{2}$ or $\frac{2}{3}$. (Perhaps there are two rifles available to her, one with $p = \frac{1}{2}$ and the other with $p = \frac{2}{3}$, and the actual rifle used is determined by the flip of a fair coin.) To determine the value of $p$ you conduct an experiment; you ask her to shoot at the target 300 times. If $p = \frac{1}{2}$, then you expect about $300 \times \frac{1}{2} = 150$ hits, and if $p = \frac{2}{3}$, then you expect about $300 \times \frac{2}{3} = 200$ hits. When she is done shooting, however, you see she has hit the target 175 times, exactly midway between 150 and 200 hits. So what is your decision?

This is a very simple example of what mathematicians call *hypothesis testing*; that is, you have to decide *after* the experiment which hypothesis ($p = \frac{1}{2}$ or $p = \frac{2}{3}$), that is equally likely *before* the experiment, to accept. You can use doggy math to do this by applying what is called *maximum likelihood*, which is a fancy way of saying, "Let's first calculate the probabilities of 175 hits out of 300 shots for $p = \frac{1}{2}$ and then again for $p = \frac{2}{3}$, and then decide in favor of the value that gives the larger probability." So, all you have to do is calculate the two numbers

$$P\left(\frac{1}{2}\right) = \binom{300}{175}\left(\frac{1}{2}\right)^{175}\left(\frac{1}{2}\right)^{125} \text{ and } P\left(\frac{2}{3}\right) = \binom{300}{175}\left(\frac{2}{3}\right)^{175}\left(\frac{1}{3}\right)^{125}$$

and see which is larger. (In fact, the previous example is of the same nature; we calculated the probability of the hypothesis $p = 0.8$ and found it to be too small to be credible. So we accepted what mathematicians call the *null* hypothesis: $p \neq 0.8$.)

I like to assign this as a homework problem to see how students handle the binomial coefficients. The factorials are much too large for direct calculation, and some students are stumped. Others, however, see the trick for simply avoiding the factorials by calculating the *ratio* of $P\left(\frac{2}{3}\right)$ and $P\left(\frac{1}{2}\right)$. Then the binomial coefficients cancel to give

$$\frac{P\left(\dfrac{2}{3}\right)}{P\left(\dfrac{1}{2}\right)} = \frac{\left(\dfrac{2}{3}\right)^{175}\left(\dfrac{1}{3}\right)^{125}}{\left(\dfrac{1}{2}\right)^{175}\left(\dfrac{1}{2}\right)^{125}} = \frac{2^{175}/3^{300}}{1/2^{300}} = 2^{175}\left(\frac{2}{3}\right)^{300},$$

a calculation even my cheap drugstore machine can handle. It gives $P\left(\frac{2}{3}\right) = 0.71264 \, P\left(\frac{1}{2}\right)$. Now, $P\left(\frac{1}{2}\right) + P\left(\frac{2}{3}\right) = 1$ because $p$ is either $\frac{1}{2}$ or $\frac{2}{3}$, and so $1.71264 \, P\left(\frac{1}{2}\right) = 1$ or, at last,

$$P\left(\frac{1}{2}\right) = 0.5839 \text{ and } P\left(\frac{2}{3}\right) = 0.4161.$$

Thus, $p = \frac{1}{2}$ is the maximum likelihood choice, although with probability 0.4161, it is the *wrong* choice.

These last two examples are far more interesting applications of doggy problem math than is Vos Savant's but, still, they are really just drill problems too. I think you'll find the problems that follow to be a distinct step up in their challenge. I hope you find them both instructive and fun. At least I did, and that's because I subscribe to words attributed to Beresford Parlett, an applied mathematician at the University of California at Berkeley: "Only wimps do the general case. True teachers tackle examples."

In other words, don't let abstraction, no matter how beautiful, blind you to the hard realities of the practical world. The quantum physicist Werner Heisenberg (1901–1976), winner of the 1934 Nobel prize, said this even more bluntly when, on being told that a famous mathematician had declared that "space is simply the field of linear operators," replied, "Nonsense, space is blue and birds fly through it." Keep that message in mind as you read the problems in this book.

# The Problems

## 1. How to Ask an Embarrassing Question
## (A Warm-up Question Requiring Only Arithmetic)

Suppose you are assigned the following task: You are to determine the fraction of the population that practices a certain private act (use your imagination). If you could gather a large number of randomly selected people together into a large room or auditorium, you could then simply have each person fill out an anonymous questionnaire. Since no one could be identified from such a form, people would presumably tell the truth. But suppose this is not possible, and your task is to be accomplished over time through individual encounters. It is clear that you cannot just ask people, because they may or may not answer truthfully when confronted with the embarrassing question (EQ); you can't even use the form, because now no one "gets lost in the crowd." So how can you gather accurate data? You can use the following clever technique that I found described in a medical journal article.

You, or a member of your staff, give each person in the survey a fair coin as he or she arrives at your office. You then ask the individual to briefly step into a private room.

There, all alone, the person flips the coin. If it shows tails, then the person writes the answer (YES or NO) to the EQ on a piece of paper. If, on the other hand, the coin shows heads, then the subject flips the coin a second time and writes the answer (YES or NO) to the non-EQ, "Did the coin show heads on the second toss?" After that, the person returns to your office and gives you the paper.

Now, each such piece of paper will have a single YES or NO on it, and eventually you'll have, let's say, 10,000 such pieces of paper. You do not know which question each particular person was actually answering, and yet you can now calculate the answer to your question: What percentage of the population (or, at least, of the people in the survey) practices the "private act"? Explain why this is so, and illustrate your analysis by calculating the percentage of people that practice the private act if 6,230 write YES and 3,770 write NO.

## 2. When Idiots Duel

Duelling problems are of strong interest to most people, perhaps because of their "life-or-death" nature. Here is one that is also, I think, amusing. A and B decide to duel but, being poor, they have just one gun (a six-shot revolver) and only one bullet. Being dumb, as well, this does not deter them and they agree to "duel" as follows: They will insert the lone bullet into the gun's cylinder, A will then spin the cylinder and shoot at B (who, standing inches away, is impossible to miss). If the gun doesn't fire then A will give the gun to B, who will spin the cylinder and then shoot at A. This back-and-forth duel will continue until one fool shoots the other. What is the probability that A will win? Also, how many trigger pulls will occur, on average, before somebody wins? The second question has meaning, of course, only if there is a large number of pairs of duelling idiots; for any particular pair, the duel occurs just once and lasts for a well-defined, specific number of trigger pulls.

(If A and B point the gun at themselves, then this foolish business is simply the two idiots alternately playing what is usually called "Russian roulette." A less violent— but perhaps not by much—form of the idiots' duel can be found in the 'sudden death' playoff between two pro football teams that are tied at the end of regulation play. There, A is the team that gets first possession of the ball.)

The methodical way to attack our two questions is to write out what mathematicians call the *sample space* of the experiment; that is, to write down all the possible outcomes of the duel. So, if we define two events as

$$a = \text{gun fires for A}$$
$$x = \text{gun does not fire,}$$

then the sample points for which A wins are

$$a$$
$$x \, x \, a$$
$$x \, x \, x \, x \, a$$
and so on

That is, A wins only on those sample points that are of odd length. So, the answer to the first question is simply the sum of the probabilities of all the odd *length* sample points, i.e.,

$$P(A) = \frac{1}{6} + \left(\frac{5}{6}\right)^2 \frac{1}{6} + \left(\frac{5}{6}\right)^4 \frac{1}{6} + \ldots$$

where the sum has an infinite number of terms. Each term comes from the observation that if the duel lasts for exactly

$k$ trigger pulls, then the gun did not fire on the first $k-1$ trigger-pulls and did fire on the $k$th one. The sum is easy to calculate. Just write $P(A) = \frac{1}{6} S$ where

$$S = 1 + \left(\frac{5}{6}\right)^2 + \left(\frac{5}{6}\right)^4 + \cdots .$$

So, $S$ is a geometric series, summed by the standard trick; i.e., multiply through by $\left(\frac{5}{6}\right)^2$ to get

$$\left(\frac{5}{6}\right)^2 S = \left(\frac{5}{6}\right)^2 + \left(\frac{5}{6}\right)^4 + \cdots$$

and then subtract. Thus,

$$S - \left(\frac{5}{6}\right)^2 S = 1 \text{ or } S = \frac{1}{1 - \left(\frac{5}{6}\right)^2}$$

and so

$$P(A) = \frac{1}{6} \cdot \frac{1}{1 - \left(\frac{5}{6}\right)^2} = \frac{1}{6} \cdot \frac{1}{1 - \frac{25}{36}} = \frac{1}{6} \cdot \frac{36}{36 - 25} =$$

$$\frac{6}{11} = 0.545454 \cdots .$$

$P(A)$ is greater than $\frac{1}{2}$, which means A has the better shot at winning the duel than does B; that is, of course, because A has the first shot.

For the second question, if we define $D$ to be the duration of the duel (measured in trigger-pulls), then the average or *expected* value of $D$ is given by the formula

$$E(D) = \sum_{k=1}^{\infty} k P(D = k).$$

Since $P(D = k) = \left(\frac{5}{6}\right)^{k-1} \frac{1}{6}$, then

$$E(D) = \sum_{k=1}^{\infty} k \left(\frac{5}{6}\right)^{k-1} \frac{1}{6} = \frac{1}{6}\left[1 + 2\left(\frac{5}{6}\right) + 3\left(\frac{5}{6}\right)^2 + 4\left(\frac{5}{6}\right)^3 + \ldots\right].$$

Or, $E(D) = \frac{1}{6} S$, where now

$$S = 1 + 2\left(\frac{5}{6}\right) + 3\left(\frac{5}{6}\right)^2 + 4\left(\frac{5}{6}\right)^3 + \ldots.$$

This sum may look nasty to do because it is not a geometric series (with the increasing coefficient in each new term), but that is not the case. We can sum $S$ using the same trick as before. So, multiplying through by $\frac{5}{6}$, we have

$$\frac{5}{6} S = \frac{5}{6} + 2\left(\frac{5}{6}\right)^2 + 3\left(\frac{5}{6}\right)^3 + \ldots$$

and so

$$S - \frac{5}{6} S = \frac{1}{6} S = 1 + \left(\frac{5}{6}\right) + \left(\frac{5}{6}\right)^2 + \left(\frac{5}{6}\right)^3 + \ldots$$

Now, the sum on the right-hand side is a geometric series. Using the multiply-and-subtract trick one more time (I'll let you do it), we get $\frac{1}{6} S = 6$. So, $E(D) = \frac{1}{6} S = 6$.

On average, then, a duel will last for six trigger pulls. Since this is just an average, however, then some duels will be shorter (perhaps just one trigger pull) and others will be much longer. To study the magnitude of the variation in the durations of many duels we can use a computer simulation. The MATLAB program **idiots1.m** (Program 1 in the MATLAB programs at the end of this book) simulates 10,000 duels and keeps track of who wins and how long each duel lasts. The heart of the simulation is MATLAB's internal random number generator named **rand**. As with every high-level scientific programming language I am familiar with, the generator returns a number from a distribution that is uniformly distributed from 0 to 1 each time **rand** is called. (This is probably the right time to take a look at the historical essay at the end of the book, which has some discussion of random number generators in general and of MATLAB's generator in particular.) So, to simulate the firing of a six-shot revolver with just one bullet in the cylinder, simply call **rand** and see if the returned number is between 0 and $\frac{1}{6}$ (gun fires) or greater than $\frac{1}{6}$ (gun does not fire). The cylinder spin before each trigger pull makes the probability that the gun fires with each trigger pull $\frac{1}{6}$, and is also independent of all previous trigger pulls.

When I ran **idiots1.m** five times, for example, here is what was produced:

| Run | P(A) | Average Number of Trigger Pulls |
| --- | --- | --- |
| 1 | 0.5517 | 6.0089 |
| 2 | 0.5508 | 5.9424 |
| 3 | 0.5496 | 6.097 |
| 4 | 0.5492 | 5.917 |
| 5 | 0.5513 | 5.9285 |

These results are in fairly good agreement with the theoretical answers, although the estimates for $P(A)$ are a bit on the high side. It is interesting to note, from Figure 2.1, that a significant number of duels lasted longer than ten trigger pulls, and a few were longer than twenty trigger pulls.

Now, here's your problem. Our two idiots have decided, after reading the preceding analysis, to use a different procedure. They still have just one gun and one bullet but now, with each exchange of the gun, they get one addi-

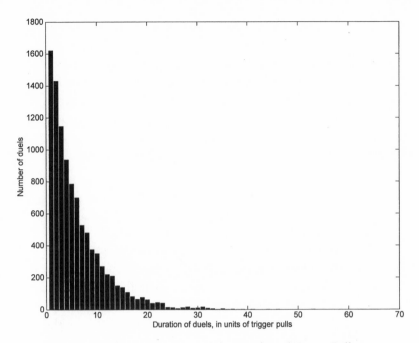

F*igure* 2.1  Relative Frequency of the Number of Trigger Pulls per Duel

tional trigger pull. That is, A puts the bullet in the cylinder, spins it and then shoots at B. If the gun doesn't fire, A gives the gun to B, who spins the cylinder and then shoots at A. If the gun doesn't fire, B spins the cylinder again and gets a second try. If the gun still doesn't fire, B gives the gun to A, who gets a maximum of three trigger pulls (with a spin of the cylinder between pulls), and so on. Calculate $P(A)$ theoretically, as well as the average number of trigger pulls over many such duels. (Hint: The second question actually requires no calculation; just think of the physics of the duel.)

A final comment: I have used these duelling problems in my classes for years, never suspecting that truth is indeed stranger than fiction. After all, only *idiots* would actually enter into such foolish competitions, right? Well, I was wrong, as you can read for yourself in the book by Kevin McAleer, *Dueling: The Cult of Honor in Fin-de-siècle Germany*, Princeton University Press, 1994. It appears that (in Germany, at least) there was no shortage of real "duelling idiots" who *took turns* shooting at each other.

## 3. Will the Light Bulb Glow?

Imagine $n$ switches in series to form a row of switches. Then, put $n$ such rows in parallel to give a sheet of $n^2$ switches. Then put $n$ such sheets in series, as shown in Figure 3.1. Imagine that each of the $n^3$ switches is, independently, either closed with probability $p$ or open (as each is shown in the figure) with probability $1 - p$. What is the probability $P_1(n, p)$ that the lamp glows?

This problem was originally given to electrical engineering students, who needed no further explanation (but not everybody is an EE.) So, the one physical concept you need to know is that for the bulb to glow there must be *at least one* complete path from the positive ( + ) terminal of the

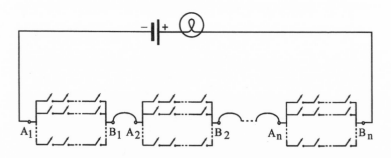

Figure 3.1  Sheets of Switches in Series

battery, through the bulb, and back to the negative $(-)$ terminal of the battery.

As a variation on this problem, imagine that the $n$ sheets are connected in parallel, rather than in series. (Parallel means to connect together the negative terminal of the battery and points $A_1$, $A_2$, . . . , $A_n$, and then connect together the positive terminal of the battery and points $B_1$, $B_2$, . . . , $B_n$). What is the probability $P_2(n, p)$ that the lamp will glow? Use a computer to study how $P_1$ and $P_2$ behave numerically as functions of $n$ and $p$.

A little twist on this question converts it into one about what is called a *random process*; i.e., a problem in which time plays a central role. This is generally not a topic in a first course in probability, except perhaps in the second semester of a year-long course, but with a little help to get you started, you should be able to do this problem. Imagine that you have a circuit path as shown in Figure 3.2, consisting of just three series-connected modules of parallel-connected switches. Imagine further that each of the mod-

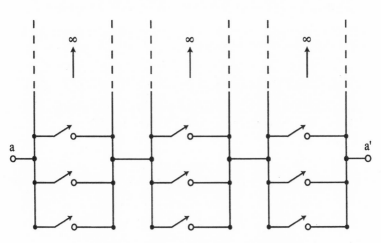

Figure 3.2   Sheets of Switches in Parallel

ules has an arbitrarily large number of switches, i.e., $n \to \infty$ where $n$ is the number of parallel switches in a module. Initially, at time $t = 0$, all the switches are open. Then, once each microsecond, one of just two events occurs: Either nothing happens with probability 0.97, or at random with probability 0.03, exactly one switch (somewhere) closes and stays closed. So, sooner or later, as the modules evolve through time, there will come a time when, for the first time, at least one switch in each module will be closed and the electrical path between terminals a and a' will be completed and the bulb will glow. Your problem is to plot the probability of the bulb's glowing, as a function of time (in units of microseconds).

As a hint to get you started, define the *state* of the path aa' at any time $t$ to be the number of modules that have at least one switch closed. (Notice that to have a completed path we need only one switch closed in each module; more than one switch closed in any given module doesn't affect the time at which the path is completed.) At time $t = 0$, then, the path is in state 0 and the path is completed for the first time as it enters state 3. So, if we write $p_k(t)$ as the probability the path is in state $k$ at time $t$, then the answer to your problem is the plot of $p_3(t)$. To do this, first explain why the following difference equations are valid (see Problems 10 and 19 for more on difference equations in probability); then, use a computer to find $p_3(t)$ from them and from the initial probabilities $p_0(0) = 1$, $p_1(0) = p_2(0) = p_3(0) = 0$:

$$p_0(t+1) = 0.97\, p_0(t)$$
$$p_1(t+1) = 0.03\, p_0(t) + 0.98\, p_1(t)$$
$$p_2(t+1) = 0.02\, p_1(t) + 0.99\, p_2(t)$$
$$p_3(t+1) = 0.01\, p_2(t) + p_3(t).$$

Or, in matrix form,

$$[p_0(t+1)\ p_1(t+1)\ p_2(t+1)\ p_3(t+1)] =$$

$$[p_0(t)\ p_1(t)\ p_2(t)\ p_3(t)] \begin{bmatrix} 0.97 & .03 & 0 & 0 \\ 0 & .98 & .02 & 0 \\ 0 & 0 & .99 & .01 \\ 0 & 0 & 0 & 1 \end{bmatrix}$$

with $[p_0(0)\ p_1(0)\ p_2(0)\ p_3(0)] = [\ 1\ 0\ 0\ 0]$. Or, even more compactly,

$$\underline{p}(t+1) = \underline{p}(t)\ \underline{P}\ ,\quad \underline{p}(0) =$$
$$[1\ \ 0\ \ 0\ \ 0],$$

where $p(t)$ is called the *state vector* of the path and $P$ (the $4 \times 4$ matrix) is the *state transition* probability matrix of the path. Notice that each row of $P$ sums to one. Can you explain why? Probably, if you think about what each of the $P(i, j)$ probabilities denotes.

This last problem is a very simple example of what is called a *Markov chain*, after the great Russian mathematician Andrei Andreevich Markov (1856–1922), who pioneered this extremely important branch of probability with a paper in 1906. Markov must have been quite an interesting fellow; he was well-known for speaking his mind without regard to political consequences. When the Czarist government organized a celebration in 1913, for example, in honor of the three-hundredth year anniversary of the house of Romanov, Markov responded by counter-organizing a celebration of the two-hundredth year anniversary of the publication of the weak law of large numbers in James Bernoulli's seminal *Ars Conjectandi* (1713). Luckily for Markov, the authorities dismissed his loudly expressed liberal views as simply those of an eccentric academic crank, or his career would surely have ended long before his natural death (after much suffering, following an eye operation). Problem 8 has some more discussion of Markov chains.

## 4. The Underdog and the World Series

The mathematics of the doggy problem has applications far removed from dogs and missiles. For example, the World Series is, as a first and crude approximation, a Bernoulli sequence of trials. Each trial is, of course, the playing of an individual game. Suppose the stronger team has probability p ($> \frac{1}{2}$, of course) of winning any particular game. What is $P(p)$, the probability that the weaker team wins the series? (For those who don't follow baseball, the modern World Series is a best-of-seven competition; i.e., the first team to win four games wins the series.) In addition, find an expression for $D(p)$, the average duration (in games) of the World Series as a function of $p$, and use the following actual historical data to estimate $p$ (this is very crude, as it assumes $p$ has been constant for nearly a century, but this is for fun).

### Historical Data

From 1905 through 1999 there have been 91 World Series played by the modern rules. The first series, in 1903, was best-of-nine and so has been excluded, as were the series for 1919, 1920, and 1921, for the same reason. The series for 1904 was simply not scheduled, and the 1994 series was canceled because of a players' strike.

| Duration of World Series | Number of Times |
| --- | --- |
| 4 games | 17 |
| 5 games | 20 |
| 6 games | 21 |
| 7 games | 33 |

# 5. The Curious Case of the Snowy Birthdays

Probably the two most famous numbers in mathematics beyond arithmetic are $\pi$ and $e$. Both are irrational (neither is equal to a ratio of integers), so their decimal expansions never repeat. Both can be computed with any desired degree of precision, however; $\pi$ has been computed to several billion decimal places and $e$ to not so many (but still a lot). There are many famous formulas for performing such calculations, for example

$$\pi = 4\left(1 - \frac{1}{3} + \frac{1}{5} - \frac{1}{7} + \frac{1}{9} - \cdots\right)$$

and

$$e = \frac{1}{0!} + \frac{1}{1!} + \frac{1}{2!} + \frac{1}{3!} + \cdots.$$

Such exact formulas are not the only way to estimate $\pi$ and $e$, however; there are random techniques, too. One well-known approach to experimentally estimate $\pi$ is often presented in probability textbooks, based on the so-called Buffon needle tossing experiment (after Georges-Louis Leclerc (1707–1788), who became the Comte de Buffon). First proposed in a now-lost 1733 paper, and again in 1777 in his "Essai d'Arithmétique Morale," it is now too well-known to be suitable as a challenge problem; therefore, I will simply describe it and give the answers (there are two answers, with textbooks usually discussing only the simpler one). Imagine a surface upon which are ruled straight, parallel lines spaced distance $d$ apart. If a short needle of length $\ell$ ("short" means $\ell < d$) is tossed randomly onto the surface, then the probability that the needle intersects a line is $2\ell/\pi d$ (see just about any modern undergraduate probability textbook for how to derive this result). The case for a long needle $(\ell > d)$ is actually no more difficult to analyze, although I have not seen it done in any of the textbooks I've looked at; the result is the somewhat more complicated expression

$$1 - \frac{2}{\pi d}\left[\sin^{-1}\left(\frac{d}{\ell}\right) + \ell\left\{\sqrt{1 - \left(\frac{d}{\ell}\right)^2} - 1\right\}\right].$$

Notice that if $\ell = d$, the two expressions give the same probability of $2/\pi$, and that as $\ell \to \infty$, the probability of a long needle intersecting at least one line approaches 1 (which makes obvious physical sense, since as $\ell \to \infty$, the only way it would not intersect any lines is to land exactly parallel to the ruled lines, an event with a vanishingly small probability).

This physical process has become famous as a way to estimate the value of $\pi$ by performing a computer simulation (see the end of the solution for Problem 11). This can be done by letting $y = 0$ and $y = d$ be the equations of the two adjacent lines between which one end of the needle lands, letting $Y$ be a random number between 0 and $d$ that represents the ordinate of that end of the needle, and letting $\theta$ be a random number between 0 and $2\pi$ radians that represents the angle the needle makes with the horizontal. It is then a simple problem in trigonometry to calculate the ordinate of the other end of the needle (if it is greater than $d$ or less than 0, then the needle crosses a line). Finally, simply repeat the above steps many times and calculate the ratio of tosses that result in the needle crossing a line to the total number of tosses. The logical flaw in such a simulation, of course, is in the generation of values for $\theta$; we need to already know the value of $\pi$, the very thing we are trying to estimate. Such a simulation is circular. What we

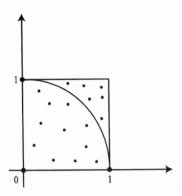

Figure 5.1   Results of Random Dart Throws

should properly simulate is a physical process that does not require an a priori knowledge of $\pi$.

For example, Figure 5.1 shows a quarter-circle with unit radius bounded by a square. Suppose we "throw darts" at the square at random; i.e., suppose that we generate $N$ random pairs of numbers with each pair representing the $x$, $y$ coordinates of where a dart lands in the square. If all the random numbers are independent, then any particular dart is as likely to land in one small patch of the square as in any other small patch of equal area. So, if we keep track of the number of darts that land in the square and below the circular arc, then we expect the ratio of this number of darts to the total number of darts to be equal to the ratio of the quarter-circle's area to the square's area. That is, we expect that ratio to be $\pi/4$. The MATLAB program **pisim.m** (program 2) performs a simulation of this process for $N = 10,000$ darts; running it several times gave the following results:

| Run | Estimate for $\pi$ |
|-----|--------------------|
| 1   | 3.1500             |
| 2   | 3.1336             |
| 3   | 3.1480             |
| 4   | 3.1280             |
| 5   | 3.1396             |

None of these individual results is particularly good, but their average value (3.13984) is within 0.06 percent of $\pi$. And no a priori knowledge of $\pi$ was required.

A much less-discussed problem in textbooks—indeed, I haven't seen it discussed in *any* textbook—is that of how to estimate $e$ by a similar tossing of darts. Does that mean there is no physical process analogous to the one just used to estimate $\pi$? No. There are, in fact, many such processes. For example, imagine that the interval 0 to 1 is divided into $N$ equal subintervals. Then, generate $N$ random numbers uniform over 0 to 1, and keep track of which subinterval each number falls into. The probability that any one of the numbers will fall into a particular subinterval is $p = 1/N$. The probability that exactly $n$ of the numbers falls into a particular subinterval is, from the doggy math/Bernoulli trials discussion in the introduction, given by

$$P(n) = \binom{N}{n} p^n (1-p)^{N-n} = \binom{N}{n} \left(\frac{1}{N}\right)^n \left(1 - \frac{1}{N}\right)^{N-n}.$$

In particular, for $n = 0$ we have $P(0) = \left(1 - \frac{1}{N}\right)^N$, i.e., this is the probability of any given subinterval receiving none of the N random numbers. But, by the definition from freshman calculus of

$$e^x = \lim_{N \to \infty} \left(1 + \frac{x}{N}\right)^N,$$

we see that for a large $N$ we can write $P(0) \approx e^{-1}$.

So, imagine that we perform the following experiment. First, generate $N$ random numbers uniform from 0 to 1, and put them into $N$ "bins" of equal width (this is thus a one-dimensional random tossing of darts). Next, count the number of bins, $Z$, that receive none of the random numbers. Thus, $Z/N$ estimates $P(0)$ and so

$$e \approx \frac{N}{Z}.$$

Writing a computer simulation of this experiment is the first part of your assignment for this problem (use $N = 100$, 1,000, and 5,000).

For the next part of this problem, consider the following completely different random process. Calculate the random sum $S = \sum_{i=1}^{N} X_i$ where the $X_i$ are independent and uniform from 0 to 1. The value of $N$ is the first number of terms such that $S \geq 1$; since the $X_i$ are random, then clearly $N$ is a random variable, as well. The question to be answered is: What is the average number of terms required to first achieve $S \geq 1$; i.e., what is $E(N)$? Obviously, $N$ must be at least 1 (although $N = 1$ to achieve $S = 1$ occurs with probability zero, this is not impossible). $N$ will almost always be 2 or more when you form $S$. Try to guess, from your numerical result, the exact theoretical answer (as a practical hint, don't forget the title of this problem). Can you derive the theoretical answer? As a technical hint for how to do it, consider the following amusing problem, with a solution that can be just slightly modified to find $E(N)$.

Suppose it snows, as it almost surely will for Daphne Tams

on her next birthday, and that we call the amount of snow $S_0$. (If $S_0 = 0$, then it doesn't snow.) How many birthdays will Daphne have to wait, on average, until she has a birthday on which the snowfall exceeds $S_0$? Let $S_n$, $n \geq 1$, be the snowfall $n$ birthdays after her next birthday. We want to calculate $E(N)$, where $n = N$ is the subscript of the first value of $S_n > S_0$. Now, for any particular value of $n$, we can write

$$P(N > n) = P(S_0 > S_1, S_0 > S_2, \ldots, S_0 > S_n)$$

because if $N > n$ then on none of the $n$ birthdays after her next birthday will the snowfall exceed $S_0$. That is, of all the values $S_0, S_1, S_2, \ldots, S_n$, the value $S_0$ is the maximum. So,

$$P(N > n) = P[\text{maximum } (S_0, S_1, S_2, \ldots, S_n) = S_0].$$

However, since there is nothing special (mathematically) about the snowfall on any of Daphne's birthdays, then it is just as likely for any of the $S_i$ to be the maximum as it is for any other, i.e.,

$$P(N > n) = \frac{1}{n+1}.$$

The crucial observation at this point is that

$$P(N = n) = P(N > n - 1) - P(N > n),$$

and so

$$P(N = n) = \frac{1}{n} - \frac{1}{n+1} = \frac{1}{n(n+1)}.$$

But now we have just what we need because, by definition,

$$E(N) = \sum_{n=1}^{\infty} nP(N = n) = \sum_{n=1}^{\infty} \frac{1}{n+1} = \infty.$$

The last sum blows up because, of course, it is just the famous harmonic series (less the first term). This surprising result says that Daphne will, on average, have to wait *forever* for a birthday that will be snowier than her next birthday. This is surprising because $P(N > n) = \frac{1}{n+1}$ says, for example, that the probability that she will have to wait longer than nineteen birthdays is just 1/20, a fairly small probability. This problem, like problem 15, shows that the average value of a random variable may actually tell you nearly nothing about the probabilistic behavior of the random variable. You can use this same approach to find the theoretical answer to the random sum question.

Finally, for the last part of your assignment, here's another physical process involving both a sum and e. Take the unit interval and randomly divide it into two parts, called the left ($L$) and the right ($R$). So, if $X$ is uniform from 0 to 1, then $L = (0, X)$ and $R = (X, 1)$. Then, take $L$ and divide it into left and right parts, i.e., if we use the subscript 1 to indicate the first division, then $L_1 = (0, X_1)$ and $L_2 = (0, X_2)$, where $X_1$ is uniform from 0 to 1 and $X_2$ is uniform from 0 to $X_1$. Keep doing this; i.e., form

$$L_3 = (0, X_3), \quad X_3 \text{ uniform from 0 to } X_2,$$
$$L_4 = (0, X_4), \quad X_4 \text{ uniform from 0 to } X_3,$$
and so on.

Now, define the random sum $S = \sum_{n=1}^{\infty} X_n$; the question for you is, what is the probability distribution for S? That is, what is $P(S \leq s)$, the probability that $S \leq s$? This is not a trivial problem to do theoretically; in *The American Mathematical Monthly* of June–July 1998 (pp. 561–562), it is shown that, for the interval $0 \leq s \leq 1$, $P(S \leq s) = e^{-\gamma}s$ where $\gamma$ is Euler's constant ($\gamma = 0.57721566490153286 \ldots$). Your assignment for this part is to simulate the "slice-and-dice" process to confirm that $P(S \leq s)$ does indeed increase linearly for $0 \leq s \leq 1$. (By measuring the slope of the $P(S \leq s)$ plot you can even estimate *e*.)

# 6. When Human Flesh Begins to Fail

Sometimes you can write down a formula for the answer to a probability problem without too much effort, and yet be unable to evaluate it numerically because the number crunching simply becomes overwhelming. Modern personal computers have done a lot to make such concerns far less serious than they once were. As an example of such a problem, consider $N$ people, all independently flipping their own fair coins. If each flips his or her coin $n$ times, then what is the probability that all $N$ people get the same number of heads? Writing down a formula, as a sum, shouldn't be hard, but evaluating it for anything other than small values of $N$ and $n$ can be a Herculean task. So, having said that, solve this problem for the cases of $N = 2$, 3, and 4 people, with $n = 10$, 50, 100, and 150 flips (a total of twelve answers).

# 7. Baseball Again, and Mortal Flesh, Too

This problem combines baseball and massive numerical computation, an interesting combination not normally found on the playing field. Suppose a team has probability $p$ of winning any particular game, and that it plays $n$ independent games. This is the same assumption made in Problem 4 ("The Underdog and the World Series"), but here it is much less justifiable. In the World Series, each team plays only a single opponent, so arguing that $p$ is constant is at least somewhat plausible. In the course of a season, however, each team plays many different opponents and the probability of winning against each one clearly depends on the relative strengths of the teams. Still, with that said, I'll continue to use a constant value for $p$, as this problem will be tough enough even with our gross simplification.

After playing $n$ games, we would expect our team to have won $np$ games, although of course it might win any number from none to all. For each possible number of games won—call this number $k$—we can use the doggy math/Bernoulli trials discussion from the introduction to write the probability of winning $k$ games as

$$\binom{n}{k} p^k (1-p)^{n-k}.$$

In a regular baseball season there are 162 games for each team, so we would expect our team to win $162p$ games by season's end, and we would expect it to have won $81p$ games by midseason. Two interesting, obvious questions are, thus: What is the probability that at midseason the team has won at least $81p$ games, and what is the probability that at season's end it has won at least $162p$ games? Some students argue that these two probabilities should be equal (or at least nearly so), but this is not the case.

You can show this as follows. Define $P_n(p)$ as the probability a team wins at least $np$ games out of $n$, if it wins any particular game with probability $p$. Then, study the behavior of $P_{81}(p)/P_{162}(p)$ as a function of $p$, over the interval $0 < p < 1$. Figure 7.1 shows this ratio for the 99 values of $p$ from 0.01 to 0.99 in steps of 0.01. The plot shows that the ratio oscillates about 1, and that in fact it can deviate significantly from 1. That's interesting, but as it turns out that is not the most interesting feature of the plot.

The plot looks "choppy," an appearance I originally attributed to using an insufficient number of $p$-values. So, I thought, I'll simply use more $p$-values and get a smoother plot, e.g., use the 199 values for $p$ from 0.005 to 0.995 in steps of 0.005. This leads to your assignment. Write down

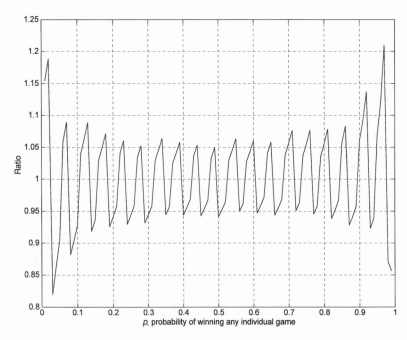

Figure 7.1 Ratio of Probabilities of Winning at Least $np$ Games Out of $n$, $n = 81/n = 162$, $p$-resolution $= 0.01$

the expression for $P_{81}(p)/P_{162}(p)$, use it to verify Figure 7.1, and then increase the $p$-resolution to see if you do get a smoother curve. Do you have an explanation for what you see?

## 8. Ball Madness

It was almost de rigueur in older probability textbooks to include at least some discussion of a class of problems that go under the general heading of "random drawings of balls out of urns." There was a logical reason for this, too; physicists have long used the basic ideas of placing balls in urns to help them visualize various complicated physical processes. Distributing electrons over various energy orbits around a central nucleus in accordance with the constraints of quantum mechanics is a balls-and-urns problem (the electrons are the balls and the orbits are the urns), for example.

Another example, which requires no knowledge of quantum mechanics, is that of gaseous diffusion. It is due to the Austrian physicist Paul Ehrenfest (1880–1933) and his Russian mathematician wife/collaborator Tatyana (1876–1964). In a 1907 paper, they imagined the following situation: Two urns (let's call them I and II) each contain $n$ balls. Initially, at time $t = 0$, all of the balls in I are black and all of the balls in II are white. (We could imagine, for example, that the white and black balls are the molecules of two different gases in a sealed chamber, initially separated by a thin membrane.) Then, at time $t = 1$ (in arbitrary units), a ball is selected at random from each urn and instantaneously placed in the other urn. (We could imagine, for example, that the membrane is punctured and the molecules are now free to move from one side of the chamber to the other.) This select-and-transfer, or exchange process, is repeated at times $t = 2, 3, \ldots$. At any given time each urn always contains $n$ balls, but only at $t = 0$ are the colors of the balls in a given urn necessarily identical.

The words "selected at random" mean that the probability of selecting a black ball from an urn containing $b$ black balls is $b/n$. At any given time, the state of both urns (which I'll call the *system*) is completely determined by specifying the number of black balls in I (or the number of white balls in II). The problem of determining the system state as a function of time is a Markov chain problem (see Problem 3), and the state transition probabilities are easy to write down. Specifically if (as in Problem 3) $P(i, j)$ is the probability that the system, in state $i$ ($i$ black balls in I) at time $t$ changes to state $j$ ($j$ black balls in I) at time $t+1$, then

$$P(i, j) = \left(\frac{n-i}{n}\right)^2 \text{ if } j = i+1 \text{ and } i < n,$$

$$P(i, j) = \frac{2i(n-i)}{n^2} \text{ if } j = i,$$

$$P(i, j) = \left(\frac{i}{n}\right)^2 \text{ if } j = i-1 \text{ and } i > 0,$$

$$P(i, j) = 0 \text{ otherwise.}$$

So, given the value of $n$ we could then just write out the state transition matrix as in Problem 3, and proceed. But it would be a huge matrix, as the number of gas molecules in a volume even just the size of a pinhead, at normal temperature and pressure, is in the many millions. Rather than using matrix arithmetic, then, let's just simulate the physics of the ball exchange process, a simulation performed by the MATLAB program **gas.m** (program 3). Figure 8.1 shows the case of $n = 50$ balls in each urn, over the time interval $0 \le t \le 600$. We can make two observations from the plot. First, the system state quickly evolves to where there are about an equal number of black and of white balls in each urn. That may even be intuitive for most people, but more subtle is that the fluctuations of the state around this nominal condition are both significant in magnitude and apparently unceasing. These fluctuations can, however, be decreased by increasing $n$, at the price of a longer time to reach the equilibrium state; Figure 8.2 demonstrates this for the case of $n = 1,000$ balls over the time interval $0 \le t \le 10,000$.

The first part of your assignment in this problem is to explain why the above state transition probabilities are correct.

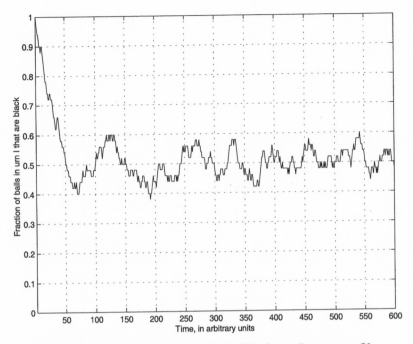

Figure 8.1 Simulation of the Ehrenfest Ball Exchange Process, $n = 50$

While not as popular among modern engineering text-book authors as they once were, balls-and-urns problems can still be entertaining and informative. And, in fact, the actual drawing of balls out of urns does still occur today in a few cases that involve a lot of money, such as some lotteries. The National Basketball Association (NBA), for example, uses the balls-out-of-an-urn process to determined the order in which teams make their picks of new players in its draft lottery. The NBA lottery is easy to understand: Four numbered balls (out of fourteen) are drawn. The balls are numbered from 1 to 14, and so the number of different combinations *without regard to the order of the drawn numbers* is

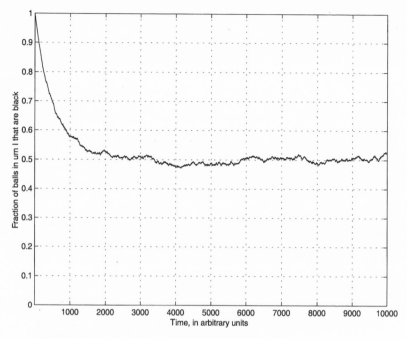

Figure 8.2 Simulation of the Ehrenfest Ball Exchange Process, $n = 1000$

$$\frac{14 \times 13 \times 12 \times 11}{4 \times 3 \times 2 \times 1} = 1,001.$$

To keep the math simple, the NBA discards one combination and the remaining one thousand combinations are assigned in various quantities to all the teams. The worse its record, the more combinations a team gets (e.g., in the 1997 draft lottery, the once-great Boston Celtics were given 200 combinations). If, on a particular drawing, the combination selected is among those held by a team, the team wins that pick.

But figuring the odds in the NBA lottery is too simple to be of interest for us; we need a tougher problem. So, with a tip of the hat to textbook tradition, here is the rest of this book's one balls-out-of-an-urn problem.

Imagine an urn with $n$ balls, identical in every respect except for being numbered from 1 through n so they can be distinguished. You draw the balls from the urn, one at a time, with replacement and a vigorous shaking after each draw. Clearly, after the $n$th draw you are certain to draw a ball that you drew earlier. You might think, however, that if $n$ is a large number, say $n = 10,000$ balls, that it would be unlikely to have a repeat drawing early in this process. When I've asked my students to use their intuition, the usual response is something like $\frac{1}{2}n$, i.e., they think it would take on the order of 5,000 draws before it would be likely to draw a previously drawn ball from 10,000 balls.

They are wrong; although astonishing, a bit of analysis will show that a repeat ball will occur, on average, in far fewer than 200 draws if $n = 10,000$ balls. That's the next part of your assignment: Derive a formula for the average number of draws before a ball is drawn for the second time, and evaluate it for the cases of $n = 10$, $n = 100$, $n = 1,000$, $n = 10,000$, and $n = 100,000$ balls. To be clear about the "before" in the last sentence, a drawing sequence of 2, 9, 7, 3, 4, 7 would, for example, be recorded as requiring five draws before the first repeat occurs (the ball numbered 7) on the sixth draw. As a related second question, what is the greatest integer $T$ of drawings for which the probability of *not* having a repetition is still greater than $\frac{1}{2}$? Once again, I think you'll be amazed at how small the answer is, even for $n$ quite large (use the same values for $n$ as in the first question).

A final word: When the Italian-born MIT mathematician and philosopher Gian-Carlo Rota (1932–1999) died, one of his obituaries noted that he was fond of saying that his considerable reputation in the field of combinatorics was based on his skill at "putting different colored marbles in different colored boxes, at seeing how many ways you can divide them." So, you see, problems like the ones treated here do indeed here have a distinguished heritage.

## 9. Who Pays for the Coffee?

A popular game of chance played perhaps a million times or more a day across the nation, whenever a lunch truck pulls into a company parking lot at break time, is "odd-person-out." If each of $N$ people desires a cup of coffee, then each one individually flips, a fair coin simultaneously with the others, to determine the one person who will pay for all $N$ cups. If all the coins but one show the same face, then the odd person out is the one who pays. If any other combination of heads and tails shows on the coins, then all $N$ people flip again. It is clear that if $N$ is large, then it might take a number of flips for an odd person out to occur. So, an important (to the lunch crowd) question is: On average, how many flips are required to get an odd person out when $N$ people play with fair coins?

This is yet another example of the common appearance of doggy math and Bernoulli trials. First, I'll calculate the probability that there is an odd person out on any particular flip. That can happen in just two ways: (1) $N-1$ people get tails and one person gets heads; (2) $N-1$ people get heads and one person gets tails. The two cases are obviously of equal probability, so I'll calculate the probability of (1) and then double it.

There are $N$ ways to pick the person who gets heads (which happens with probability $\frac{1}{2}$) and one way for the other $N-1$ people who all get tails (which happens with probability $\left(\frac{1}{2}\right)^{N-1}$). So, the probability of (1) is $N \frac{1}{2}$ $\left(\frac{1}{2}\right)^{N-1} = N\left(\frac{1}{2}\right)^N$. This is also the probability of (2). So, if we let $p$ denote the probability of an odd person out on any particular flip, then

$$p = 2N\left(\frac{1}{2}\right)^N = \frac{N}{2^{N-1}}.$$

Next, I'll compute the probability $P_k$ that it takes exactly $k$ flips to achieve an odd person out. This is a Bernoulli sequence of $k-1$ flips with no odd person out (which happens with probability $1-p$ on each such no-decision flip) followed by the $k$th flip, which does reach a decision (an odd person out occurs, with probability $p$). So,

$$P_k = (1-p)^{k-1}\, p.$$

The answer to our question then, the average number of flips it will take to get an odd person out, is

$$\sum_{k=1}^{\infty} kP_k = p\sum_{k=1}^{\infty} k(1-p)^{k-1} = pS$$

where $S$ denotes the last summation. Writing $S$ out term by term,

$$S = 1 + 2(1-p) + 3(1-p)^2 + 4(1-p)^3 + \ldots .$$

Multiplying through by $(1-p)$,

$$(1-p)\,S = (1-p) + 2(1-p)^2 + 3(1-p)^3 + \ldots .$$

Subtracting the last expression from the previous one,

$$S - (1-p)S = pS = 1 + (1-p) + (1-p)^2 + (1-p)^3 + \ldots .$$

Now, the right-hand side is a geometric series summable as follows: write the right-hand side as

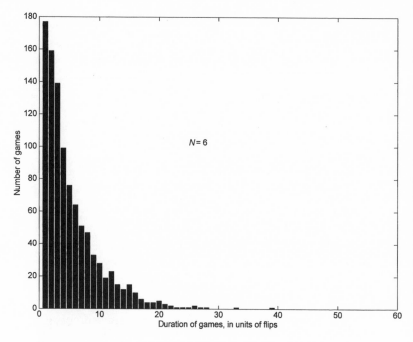

Figure 9.1   Relative Frequency of the Duration of "Odd-Person-Out"

$$T = 1 + (1-p) + (1-p)^2 + (1-p)^3 + \dots .$$

Thus, $\quad (1-p)\, T = (1-p) + (1-p)^2 + (1-p)^3 + \dots .$

and so $\quad T - (1-p)\, T = pT = 1$ or, $T = 1/p$.

Now, $pS = T$ and so $S = 1/p^2$. But the average number of flips to achieve an odd person out is $pS = 1/p$, so our answer is $1/p = 2^{N-1}/N$ flips.

To check this theoretical result, the MATLAB program **odd.m** simulates 1,000 games of odd-person-out for any desired value of $N$. It keeps track of how long each simulated game lasts (in units of flips) and computes the average duration of the games. As Figure 9.1 shows, there can be games that last significantly longer than the average duration. The following table compares the theory with the simulation.

| | Average duration (flips) | | |
| --- | --- | --- | --- |
| N | Theory | odd.m (three simulations) | |
| 3 | 1.333 | 1.305 | 1.336 | 1.308 |
| 4 | 2 | 1.981 | 2.045 | 2.03 |
| 5 | 3.2 | 3.183 | 3.121 | 3.082 |
| 6 | 5.333 | 5.475 | 5.239 | 5.29 |

With all that done, here's your assignment. Suppose $N - 1$ people have fair coins and the Nth person has a biased coin, i.e., a coin that shows heads with probability $q$ and tails with probability $1 - q$. How does this change the theoretical result? Modify **odd.m** (program 4) to simulate this new situation, and use it to confirm your answer.

## 10. The Chess Champ versus the Gunslinger

Whenever somebody is the "best" at just about anything, the challenge is there for somebody else to try to beat him or her. In the Old West, it was often about who was the best in a shootout; in today's more civilized (?) times it might be about who is the best chess player. (In fact, it appears as if it is now IBM's Deep Blue computer, which defeated world champion Garry Kasparov in 1997.) The chess challenge leads, in fact, to a fascinating problem that I believe is still unsolved as I write, where by "solved" I mean in terms of closed-form analytical expressions. Here it is.

Suppose the world chess champion plays a challenge match of $N$ games with the latest young "gunslinger." The champ has probability $p$ of winning any particular game, and there is probability $q$ that the game ends in a draw. In chess, a win is worth one point and a draw is worth half a point to each player. At the end of the Nth game (N is fixed at the start of the match), the winner is simply the player with the most points. Chess is played with white and black pieces, with the opponents alternating as to who plays white (and thus gets the first move). Since having the first move is of considerable advantage, N is even in match play, to give both players the same number of times with the white pieces. (In the Kasparov/Deep Blue 1997 match, with $N = 6$, Kasparov won one game, Deep Blue won two, and there were three draws, for a final score of $3\frac{1}{2}$ to $2\frac{1}{2}$.) So, here's your general assignment.

What is the probability that the champ wins the match, and what is the probability that the match ends in a tie? While analytical solutions to these questions are unknown to me, a computer solution can be done. As I've done in most of the problems in this book, you could just write a Monte Carlo simulation, e.g., as in the duelling idiots problem. But since we are interested in how the answers vary with $p$ and $q$, then if we step $p$ and $q$ by increments of 0.01 over their intervals (say, $0 \le q \le 0.3$ with $0 \le p \le 1 - q$), and for each $p$, $q$ combination we simulate just 100 games (getting, at best, pretty crude estimates), there will still be a lot of simulation required. So, for this problem I suggest you take a different computer approach. Define $C(k, n)$ as the probability that the champ is ahead by $k$ points at the end of the nth game. Can you write a recurrence equation for $C(k, n)$? Can you write down *by inspection* what $C(k, k)$ is? How about $C(k, n)$ when $k > n$? With just a bit more effort, can you write down a formula for $C(0, n)$? If your answer to all four questions is yes, then you are on your way. If not, here are three extra hints.

1. An example of a recurrence: Writing a recurrence or *difference* equation involving probabilities can

often be the key to solving a problem. For example, consider the case of flipping a coin $n$ times; if heads show with probability $p$, what is the probability of getting an even number of heads? Define the answer to be $P_n$. Now, the first flip is either a head or a tail. If a head, then we need an odd number of heads on the next $n - 1$ flips, and if a tail, then we need an even number of heads on the next $n - 1$ flips. So, we have the first-order recurrence for $P_n$ of

$$P_n = p(1 - P_{n-1}) + (1 - p)P_{n-1} = (1 - 2p)P_{n-1} + p.$$

(Zero heads counts as even, so we also know $P_1 = 1 - p$.) This particular equation is not hard to solve (you can easily verify that $P_n = \frac{1}{2}[1 + (1 - 2p)^n]$ by simply substituting it into the recurrence) but the point here is that, for a numerical solution, you don't have to know how to solve the recurrence. Using the recurrence and the value for $P_1$ you can find $P_2$. Then use $P_2$ (and the recurrence) to find $P_3$, and so on. You can do the same for the $C(k, n)$. Problems 3 and 19 have more discussion about recurrence equations in probability.

2. Once you get a program written, running, and producing numbers, it is a good idea to think of some simple tests with which to validate those numbers. That is, ask yourself: Are there special cases for which you can calculate some of the probabilities asked for, and to then use as checks on what the computer is producing? Here are a couple of such cases you can use to at least partially validate your chess program.

a. Suppose $N$ is even and $q = 0$ (i.e., there are no drawn games). Then, for the match to be tied at the end, each player must have won half the games. This occurs with probability

$$\binom{N}{\frac{N}{2}} p^{N/2} (1-p)^{N/2}$$

which, for $N = 6$, is $20\, p^3 (1-p)^3$.

b. Suppose $N$ is odd and $q = 0$. Then, there is no way the match could be tied at the end; i.e., the probability of a tied match is zero for all $p$.

3. This problem, which I think astonishingly deceptive in its apparent lack of difficulty, has recently yielded to theory just a little bit. In a paper published in the November 1998 issue of *The American Mathematical Monthly* ("The Probability of a Tie in an N-Game Match," vol. 105 pp. 844–846), J. Marshall Ash derived an asymptotic formula for the probability of a tie in a chess match if $p = q = \frac{1}{3}$. For this very special case, Ash showed that the probability of a tie limits, as the number of games N in a match becomes arbitrarily large, to $\sqrt{\frac{3}{4\pi N}}$. This result is particularly interesting because it suggests, even with $N$ even, that the probability of a tied match approaches zero (some observers of chess have felt that an even $N$ tends to favor a tied match).

Here's your specific assignment: Write a program that calculates and plots the probabilities that an N-game match ends in a tie, is won by the champ, or is won by the challenger. Run your program for $N = 6$ and $q = 0.1$ (and so your plots will be for $0 \leq p \leq 0.9$). Then, to partially vali-date your program, run it for the $N = 6$, $q = 0$ and $N = 7$, $q = 0$ cases. Do you get what you expect to get? Also, use Ash's special result to confirm that your program agrees with the theory in the special case of $p = q = \frac{1}{3}$, for large $N$.

## 11. A Different Slice of Probabilistic Pi

This is a challenge problem in what is called *geometric prob-ability*. Such problems occur when multiple independent random variables are uniformly distributed. That reduces the problem of calculating the probability of an event to the calculation of the ratio of certain areas in a sample space diagram, which can be done using geometry; hence the name. (Problem 14 has some additional examples of this procedure.) Such problems are generally limited in their applicability (the underlying assumption of unifor-mity is fairly restrictive), but they can be both entertaining and highly subtle. More sophisticated techniques are often much harder to apply, and even then the problem may not yield easily. Every modern probability textbook I know of includes at least one example (often, the Buffon needle problem) of this type of problem. Here's one I have not seen in any textbook. The answer, as with the Buffon prob-lem, involves $\pi$.

Imagine a circular tabletop of radius $r$. Imagine further that we toss, at random, a thin steel needle of uniform den-sity and with length $2a$ (where $r \geq a$, i.e., the needle is no longer than the diameter of the tabletop) onto the table-top. We do this a very large number of times and, among all the tosses where the needle comes to rest on the table-top, we observe (1) how many times one end of the needle sticks out over the edge of the top; (2) how many times both ends of the needle stick out over the edge; and (3) how many times neither end sticks out over the edge. From these numbers we could estimate the probabilities of these three events. Rather than performing this boring experiment, however, you are to find theoretical expressions for the three probabilities, and thus show they are functions only of the ratio $a/r$. Plot your expressions for $0 \leq a/r \leq 1$. As a hint to get you started, make the physical observation that for the needle to come to rest on the tabletop, the midpoint of the needle must be on the tabletop. Assume that the midpoint is as likely to end up in any small patch of area as in any other small patch of equal area.

## 12. When Negativity Is a No-No

This is the one challenge problem in the book that was created by my students and solved by me, rather than the other way around. Just before the final exam in course EE647 during the spring semester of 1993, late one Friday afternoon, a group of students knocked on my office door. They were several of the better students in the class, and had been making up their own problems as practice study questions. They knew they had incorrectly solved one problem they had created, but couldn't understand where they had gone wrong. They knew they were wrong because their solution called for a probability density function that could be negative over part of its interval. That is impossible, of course, because that says the probability that the associated random variable lies in that interval would be less than zero (and all probabilities are strictly from 0 to 1, with no exceptions).

I promised the students that I would take a look at their problem, expecting not to have any trouble with it. After working on it all weekend, I finally did manage to get the correct density function, but it was just a bit tougher than I originally thought. Here it is, for you to try your hand with it. Suppose X and Y are independent random variables, and both are uniformly distributed from 0 to 1. What is the probability density function of Z, where

$$Z = \frac{X}{X - Y}?$$

That is, what is $f_Z(z)$ where
$$\text{Probability }(a \leq Z \leq b) = \int_a^b f_Z(z)dz?$$

## 13. The Power of Randomness

*Suppose* X and Y are independent random variables, and that both are uniform from 0 to 1. It is common in probability textbooks to see examples of how to calculate the probability density functions (pdf) for such combinations as $Z = X + Y$, $Z = XY$, or $Z = X/Y$. One combination I haven't seen in a textbook, however, is $Z = X^Y$.

So, what is the pdf for this Z, a random variable raised to a random power? That is, what is $f_Z(z)$ such that

$$\text{Probability }(a \leq Z \leq b) = \int_a^b f_Z(z)dz?$$

This calculation is actually pretty straightforward, for the most part, but it does offer a couple of subtle opportunities for the unwary to fall into a black hole. So be wary!

In addition, the answer comes out in terms of a tabulated, yet somewhat unfamiliar, function (at least that is so for most undergraduates). You may, therefore, find it helpful to know that the *exponential integral* function is defined as

$$Ei(x) = \int\limits_{x}^{\infty} \frac{e^{-u}}{u}\, du.$$

MATLAB provides this integral as a built-in function (**expint**), so it is easy to create a plot of the pdf of $Z$. As a check of your theoretical answer, use MATLAB's random number generator to produce 20,000 values of $X^Y$ and print a histogram of them. Do your two plots resemble each other?

## 14. The Random Radio

In 1883, when interest was at a fever pitch for finding a way to confirm the prediction by the Scottish mathematical physicist James Clerk Maxwell (1831–1879) of the existence of electromagnetic waves traveling through empty space at the speed of light, the Irish physicist George Francis FitzGerald (1851–1901) made a suggestion. In one of the shortest scientific papers ever published (a mere one paragraph), he described a charged capacitor being allowed to suddenly discharge through an inductive circuit (one with lots of wire wrapped around an iron core, for example). If the component values are properly chosen, then the discharge is in the form of a high-frequency oscillation. Perhaps, suggested FitzGerald, these oscillations would be sufficient to launch electromagnetic energy into space and thus create Maxwell's waves. Four years later, the German physicist Heinrich Hertz (1857–1894) did just that, and then others turned the physics into what soon became the worldwide sensation of commercial radio.

Without going through the details of FitzGerald's math, what his work boiled down to is essentially the following: Given an inductor (e.g., a coil of wire), a capacitor (a component that stores electricity and that, even in FitzGerald's day, could be easily constructed from nothing more than tinfoil and a glass jar), and the inherent resistance of the circuit itself, it is not hard to show that the discharge current decays exponentially with time. The way the exponential decays occur can be either monotonic or oscillatory, however, and which decay mode occurs depends on the component values. It turns out that what determines the mode is a particular quadratic equation, with coefficients determined by the component values. If the quadratic has real solutions then one mode occurs, and if the solutions are complex then the other mode occurs.

So, imagine that an early radio experimenter had several boxes of components on his workbench: one containing an assortment of capacitors, another containing various inductors, and a third with various resistors. If he simply grabbed components at random and connected them into Fitz-Gerald's circuit, then the resulting current might or might not oscillate. This simple illustration thus leads us to the following pure mathematics question in probability.

Consider the partially random quadratic equation $x^2 + Bx + C = 0$, where $B$ and $C$ are independent random variables uniform from 0 to 1. (I call this partially random since the coefficient of $x^2$ is not random.) What is the probability that the solution to the partially random quadratic is real? This is a classic problem in geometric probability, with variations on it appearing in practically all introductory probability theory textbooks. It is a straightforward question to answer. From the quadratic formula, we have

$$x = \frac{-B \pm \sqrt{B^2 - 4C}}{2},$$

which tells us that $x$ is real if and only if $B^2 \geq 4C$. That is,

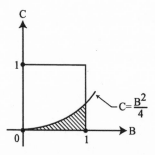

Figure 14.1   Real Roots of Quadratic Equations

we wish to calculate the probability that $C \leq \frac{1}{4} B^2$. The sample space for $B$ and $C$, on which this inequality is sketched, looks like Figure 14.1, where the shaded region represents the collection of all pairs of values for $B$ and $C$ that result in real roots.

So, the answer to the question is the probability of the shaded region. Since $B$ and $C$ are each uniform and independent, then the probability we want is simply

$$\frac{\text{area of shaded region}}{\text{area of sample space} \ (= 1)} = \int_0^1 \frac{1}{4} B^2 \, dB = \frac{1}{4} \left( \frac{1}{3} B^3 \bigg|_0^1 \right) = \frac{1}{12}.$$

This problem is also a simple example of the kind of probability question that is very easy to answer by computer simulation, a powerful method useful for either checking a theoretical analysis or for just getting a quick numerical result without the analysis. So, using the random number generator available in MATLAB, I asked for 100,000 pairs of values for $B$ and $C$. Checking each pair to see if it satisfied $B^2 \geq 4C$, I got the answer 0.08411, which compares fairly well with $\frac{1}{12} = 0.08333. \ldots$

A surprisingly simple twist—and one I haven't seen in any textbook—to the partially random quadratic demands a more sophisticated analysis (although it remains elementary). So, consider now the totally random quadratic equation $Ax^2 + Bx + C = 0$, where $A$ is also uniform from 0 to 1 (and independent of both $B$ and $C$). What is the probability that the solution to the totally random quadratic is real? You cannot simply divide through by $A$ and declare

$$x^2 + B/A \; x + C/A = 0$$

to be a partially random quadratic with the answer computed above. First of all, $B/A$ and $C/A$ aren't uniform; and, in fact, they aren't even independent. This assertion usually astonishes students (and surprised me), who argue that since $B$ and $C$ are independent, then why would dividing both by the same value ($A$) suddenly make the ratios dependent? It does, and you are about to prove it. Here's your assignment:

1. Calculate the probability of real roots for the totally random quadratic without making any independence assumptions other than that $A$, $B$ and $C$ are all independent.

2. Redo the problem with the assumption that $B/A$ and $C/A$ are independent and show that the resulting answer does not agree with your first calculation. Thus, conclude that the assumption of the independence of $B/A$ and $C/A$ must be wrong.

3. Write a computer simulation for the totally random quadratic and compare the result with your two theoretical calculations.

## 15. An Inconceivable Difficulty

This is a problem that is not very hard to do mathematically, but it has an answer that surprises nearly everybody.

Before starting it, however, I'll do a closely related problem. Consider a couple beginning a family; in the interest of promoting a diverse family, they decide to have children until they have both sexes. How many children can they expect to have?

Let $p$ be the probability of having a boy, so $1 - p$ is the probability of having a girl. There are just two sequences of length $n$ (i.e., $n$ children) that are allowed by the couple's rule:

1. BB ... BG, with probability $p^{n-1}(1-p)$, $n \geq 2$,

2. GG ... GB, with probability $(1-p)^{n-1}p$, $n \geq 2$.

So the answer to our couple's question is simply

$$\sum_{n=2}^{\infty} n \left[ p^{n-1}(1-p) + (1-p)^{n-1}p \right] = (1-p) \sum_{n=2}^{\infty} np^{n-1}$$
$$+ p \sum_{n=2}^{\infty} n(1-p)^{n-1}.$$

The two sums on the right are of the form $S = \sum_{n=2}^{\infty} n\, q^{n-1}$, with $q = p$ and $q = 1 - p$, respectively. So,

$$S = 2q + 3q^2 + 4q^3 + \ldots.$$

Thus,
$$qS = 2q^2 + 3q^3 + \ldots$$
and so
$$S - qS = (1 - q)S = 2q + q^2 + q^3 + \cdots$$
$$= q + [q + q^2 + q^3 + \ldots] = q + q[1 + q + q^2 + q^3 + \ldots]$$
$$= q + q\frac{1}{1-q} = \frac{q - q^2 + q}{1 - q} = \frac{2q - q^2}{1 - q} = q\frac{2 - q}{1 - q}.$$
$$S = q\frac{2 - q}{(1 - q)^2}.$$

Thus, our answer is, for the average number of children,

$$(1-p)\, p\, \frac{2-p}{(1-p)^2} + p(1-p)\, \frac{2-(1-p)}{[1-(1-p)]^2} = \frac{1-p+p^2}{p(1-p)}.$$

This result says, for example, that if $p = \frac{1}{2}$ (boys and girls equally likely), then the average number of children will be three. It is easy to show that this is the minimum average value, and that for $p \neq \frac{1}{2}$, the average number of children will be greater. Indeed, in the special cases of $p = 0$ (only girls are born) and $p = 1$ (only boys are born), the average family will have an infinite number of children! That, of course, is simultaneously mathematically correct (an opposite-sex child will never be born) and physical nonsense (for the obvious reason of endurance!).

Now, here's your problem. Suppose our couple decides to use a different rule: They will have children until they have a child that is the same sex as the first one. How many children can they now expect to have, as a function of $p$? Does your formula give the correct answers for the special cases of $p = 0$ and $p = 1$? (Before starting your analysis, be sure you know the answers for $p = 0$ and for $p = 1$.) Write a computer simulation to confirm your answer.

## 16. The Unsinkable Tub Is Sinking! How to Find Her, Fast

Suppose, one foggy afternoon, a radio distress call is received at a Coast Guard station from a ship at sea. "This is the *Unsinkable Tub*. Am sinking near « static burst » Island. Send help, fast!" And then the ship's transmitter goes dead. Unfortunately there are two islands (call them Island #1 and Island #2) that could be the approximate location of the sinking ship; one is ten miles to the south of the station and the other is ten miles to the north. The station commander knows the *Unsinkable Tub* is a sight-seeing excursion ship and could be at either island with a load of tourists. Time is of the essence, then, and he must split his fleet of $N$ search boats and search the waters off of both islands at the same time.

Because of the time of day, the commander thinks it is more likely that the *Unsinkable Tub* is at one of the islands than it is to be at the other island; call the probability it is at Island #1 $p_1$, and the probability it is at Island #2 $p_2$. Assume that each search boat has probability $p_s$ of finding the ship if it is sent to the correct island, and that all the search boats search independently. How should the commander divide his fleet of search boats to maximize the probability of finding the sinking ship? That is, what is $n$, the number of search boats that should be assigned to Island #1?

Find a general solution, and then plot the optimal $n$ as $p_1$ varies from 0 to 1 (of course, $p_2 = 1 - p_1$) for the case of $N = 13$ search boats and $p_s = 0.2$. Repeat for $N = 40$ search boats. Then, repeat the $N = 40$ case for $p_s = 0.05$. Because of the large amount of numerical calculation required to generate these three plots, you may want to consider using a computer.

## 17. A Walk in the Garden

Consider the unit square shown in Figure 17.1. A line is drawn through the square by first choosing a point $x$ on the X-axis at random from 0 to 1, and then choosing an angle $\theta$ at random from 0 to $\pi$ radians. The length of the segment of the line that is inside the square is the random variable $L$, where obviously $0 \le L \le \sqrt{2}$. What are the probability density and distribution functions for $L$?

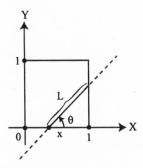

Find Probability Density and Distribution Functions for $L$

This is one of those problems that is easy to simulate on a computer, but just a bit more of a challenge to do analytically. Indeed, many more years ago than I care to think about (in 1968), I was a young member of the technical staff at the Hughes Aircraft Company in Southern California and was assigned this problem. It had appeared in the context of tracking an aircraft with radar, and I was given just forty-eight hours to come up with the answer, to be put into a new business proposal. Not yet confident of my analytical skills, I simply simulated the problem on one of the earliest timeshare systems around, in BASIC. The simulation took a while to run in those days, but I got the correct answer and I think it saved my job.

Here's a start at how to do a simulation, which will then serve as a good check of your theoretical solution. The simulation will simply generate a large number of pairs of random numbers, one for $x$ and one for $\theta$, and then use geometry to calculate the resulting $L$ for each pair. The end result will be a large number of $L$'s, from which the two desired functions can be easily constructed. There are three cases to consider, depending on which edge of the square the line cuts through. With just a bit of thought you should be able to show that:

Case 1: The line passes through the right vertical edge if

$$0 \leq \theta \leq \tan^{-1}\left(\frac{1}{1-x}\right), \text{ and } L = \frac{1-x}{\cos(\theta)}.$$

Case 2: The line passes through the top edge if

$$\tan^{-1}\left(\frac{1}{1-x}\right) < \theta \leq \pi - \tan^{-1}\left(\frac{1}{x}\right), \text{ and } L = \frac{1}{\sin(\theta)}.$$

Case 3: The line passes through the left vertical edge if

$$\pi - \tan^{-1}\left(\frac{1}{x}\right) < \theta \leq \pi, \text{ and } L = -\frac{x}{\cos(\theta)}.$$

So, here's what you are to do:

1. Write a computer simulation that generates 100,000 random paths across the unit square and then plots the density and distribution functions for $L$. You should see something "interesting" happen at $L = 1$.

2. Find the density and distribution functions for $L$ analytically, and compare them to the simulation results. Hint: Think conditional distribution function and use the theorem of total probability.

As an amusing footnote to this problem, five years after solving this problem on a computer, I found myself spending a postdoctoral year at the Naval Research Laboratory in Washington, D.C. I quickly fell into the habit of walking down to the lab's library (set near the bank of the Potomac River) at lunch to eat and read an interesting journal, picked at random. One day I happened to pick a volume of the *Journal of Applied Probability* from the shelf and flipped it open—and there was the solution to my problem! (See the paper by Maurice Horowitz, "Probability of Random Paths Across Elementary Geometrical Shapes," vol. 2, pp. 1965, 169–177.) Horowitz had published the answer three years before I had simulated the problem. What, do you think, is the probability of my finding his paper the way I did?

## 18. Two Flies Stuck on a Piece of Flypaper— How Far Apart?

Imagine that two points are placed at random, independently, inside a unit square. That is, if the coordinates of the two points are given by $(X_1, Y_1)$ and $(X_2, Y_2)$, then $X_1$, $X_2$, $Y_1$, and $Y_2$ are all uniform over to 1, and are all independent. As a physical example, think of two flies that land, at random, on a square piece of sticky flypaper. The Pythagorean distance (squared) between the two flies is given by $Z = (X_1 - X_2)^2 + (Y_1 - Y_2)^2$. It is geometrically obvious that $0 \leq Z \leq 2$. Find an expression for the probability that $Z \leq t$ where $0 \leq t \leq 2$.

Before doing the analysis, write down your intuitive guess for $P(Z \geq 1)$. Are you surprised by the actual value? Does a computer simulation agree with your formula?

Now, as a little twist to this problem, suppose the flypaper is in the shape of a circle with the same area; i.e., it has radius $1/\sqrt{\pi}$. So, $0 \leq Z \leq \frac{4}{\pi}$ where, once again, $Z$ is the Pythagorean distance squared between the two flies. Write a simulation, for this case of two flies landing at random on the circle, that estimates $P(Z \geq 1)$. Be sure to clearly state your assumptions of what it means to "land at random" on the circular flypaper.

## 19. The Blind Spider and the Fly

One of the mainstays of probability theory, particularly beloved by textbook writers, is the so-called "random walk." Random walks, so named in a 1921 paper by the Hungarian mathematician George Pólya (1887–1985), are usually quite easy to understand, but are also usually difficult (and often impossible) to answer without the aid of a computer. Here is the simplest version, the "unrestricted, symmetrical one-dimensional random walk." The infinite $x$-axis is marked off at the integers, $0 \pm 1, \pm 2, \ldots$, and a particle is placed at $x = 0$ at time $t = 0$. Once each second thereafter, the particle jumps one unit either to the left or to the right, with equal probability. So, at time $t = 1$ the particle is either at $x = +1$ or at $x = -1$; each possibility has probability $\frac{1}{2}$. At time $t = 2$ the particle could be at $x = +2$, $x = -2$, or back at $x = 0$, and so on. Each jump is independent of all the others.

The next step up in complexity puts restrictions on how far, in either direction, the particle can move. Such restrictions are called *barriers*, and are of two types. A *reflecting* barrier means that if the particle hits it then it must, on the next move, reverse direction and return to its previous position. An *absorbing* barrier means that if the particle hits it, the walk terminates. Some random walks, even with barriers, are sufficiently simple that one can obtain theoretical results. Consider, for example, the one-dimensional symmetrical random walk that starts at $x = 0$ and which has a reflecting barrier at $x = 0$ and an absorbing barrier at $x = n$. This is a symmetrical walk because when the particle is at $x = i$ its next jump is either to $x = i - 1$ or to $x = i + 1$ with equal probability *unless* it is at $x = 0$, where the reflecting barrier forces it to jump on its next move to $x = 1$ with probability 1. I will now derive a couple of important theoretical results about this random walk that will prove useful in your assignment.

First, let's prove that some time after the walk starts the particle is certain to be absorbed, i.e., the probability that the particle will reach position $x = n$ is 1. To do this, forget about the reflecting barrier. If anything, this decreases the probability that particle reaches $x = n$ because without that barrier we now have a potential escape path to minus infinity. Even so, we can show that the particle still reaches $x = n$ with probability 1.

Let $P(i)$ be the probability that the particle reaches $x = i$, which means $P(0) = 1$ (that's where the particle starts.) Now, there are just two ways to reach $x = i$: be at $x = i - 1$ and move to the right, or be at $x = i + 1$ and move to the left. So, we immediately have the recurrence equation

$$P(i) = \frac{1}{2} P(i - 1) + \frac{1}{2} P(i + 1).$$

There is now a very clever argument for why this says that all of the $P(i) = 1$. The recurrence equation says that $P(i)$ is the average of $P(i - 1)$ and $P(i + 1)$, which means that if we plot $P(i)$ versus $i$ then all of the $P(i)$ must lie on a straight line. In fact, it must be a horizontal line because if the line had a nonzero slope, then eventually we would get values for $P(i)$ that are outside the interval 0 to 1. Since the $P(i)$ are probabilities, that would of course be impossible. And since $P(0) = 1$ then $P(i) = 1$ for all $i$, i.e., in the unrestricted symmetrical random walk the particle is certain to visit every point on the infinite $x$-axis. So, returning to our problem with restrictive barriers that trap the particle to a finite region of the $x$-axis, it is certain that it will reach $x = n$ and be absorbed.

The next obvious question to ask is how many steps, on

average, the particle moves before it is absorbed. To answer this, let $J_k$ be the number of jumps made by the particle to move from $x = k$ to $x = k + 1$. The $J_k$ are, of course, random variables; in particular, each has an average value. So, let's write $e_k = E(J_k)$ as the average number of jumps required by the particle to move from $x = k$ to $x = k + 1$. If we write $J$ as the total number of jumps required for the particle to move from $x = 0$ to $x = n$, then

$$J = J_0 + J_1 + J_2 + \ldots + J_{n-1}$$

and so our answer is $E(J)$, i.e.,

$$E(J) = E(J_0) + E(J_1) + E(J_2) + \ldots + E(J_{n-1})$$
$$= e_0 + e_1 + e_2 + \ldots + e_{n-1}.$$

We must, therefore, find the $e$'s.

To do that, notice that we can write

$$J_k = \begin{cases} 1, \text{ with probability } \dfrac{1}{2} \\ 1 + J_{k-1} + J_k \text{, with probability } \dfrac{1}{2}. \end{cases}$$

The first half of this is easy to understand; it simply says that if the particle is at $x = k$ then, with probability $\frac{1}{2}$, it moves to $x = k + 1$ in one step. The second half is only a little more subtle. It says that the other way to move from $x = k$ to $x = k + 1$ is to first move backward one step to $x = k - 1$ and then to $x = k$ and finally to $x = k + 1$. So, taking expected values, we have

$$E(J_k) = \tfrac{1}{2} \cdot E(1) + \tfrac{1}{2} \cdot E(1 + J_{k-1} + J_k) = \tfrac{1}{2} + \tfrac{1}{2} + \tfrac{1}{2} E(J_{k-1}) + \tfrac{1}{2} E(J_k)$$

or, $e_k = 1 + \frac{1}{2} e_{k-1} + \frac{1}{2} e_k$
or, at last, $e_k = 2 + e_{k-1}$.

Now, because of the reflecting barrier at $x = 0$, we have

$e_0 = 1$. That is, if the particle is at $x = 0$ it (always) takes one jump to move to $x = 1$. So, combining this with the recursion for the $e$'s, we have

$$e_0 = 1$$
$$e_1 = 2 + e_0 = 3$$
$$e_2 = 2 + e_1 = 5$$
$$e_3 = 2 + e_2 = 7$$
$$.$$
$$.$$
$$.$$
$$e_{n-1} = 2(n-1) + 1 = 2n - 1.$$

Thus, our answer is

$$E(J) = 1 + 3 + 5 + 7 + \ldots + (2n - 1)$$
$$= [1 + 2 + 3 + 4 + 5 + 6 + 7 + \ldots + (2n - 1) + 2n]$$
$$- [2 + 4 + 6 + \ldots + 2n]$$
$$= [1 + 2 + 3 + \ldots + 2n] - 2[1 + 2 + 3 + \ldots + n].$$

From the well-known formula

$$1 + 2 + 3 + \ldots + n = \frac{n(n+1)}{2}$$

we have

$$E(J) = \frac{2n(2n+1)}{2} - 2\frac{n(n+1)}{2} =$$
$$\frac{n}{2}[2(2n+1) - 2(n+1)] = n^2.$$

That is, the average number of jumps made by the particle before it absorbed at $x = n$ increases with the square of the distance between the two barriers.

This is a pretty result, but it isn't at all difficult to construct random walks that defy theoretical analysis, and so must be attacked by computer simulation. As an example,

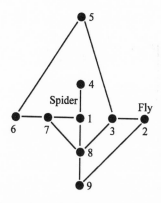

Figure 19.1 Spider's Random Walk

which will lead to your assignment, consider the following two-dimensional random walk.

Imagine an old, blind spider sitting in the middle of a somewhat tattered web, as shown in Figure 19.1. At time $t = 0$ a fly lands on one of the vertices of the web; the spider can't see the fly but he feels the thump of the landing, so he knows dinner has arrived. The fly, however, realizes his fate, dies of fright, and thereafter remains motionless on the web. Since the spider has no idea of where on the web the fly is, he begins a random walk around the web. That means that once each second he simply feels how many different strands are locally available to him at his current location and then chooses one at random with equal probability. So, at vertex 1 (his initial location) he has three possibilities, while at vertex 9 he has only two possibilities. (Spiders aren't very smart, so he may in fact actually walk right back down the very thread he just arrived on.) Now, the numbering of the web vertices is completely arbitrary, so to set a standard, let's agree to always put the spider initially at vertex 1 and the fly at vertex 2. During its walk, then, if the spider arrives at vertex 2 the walk stops, because that is the absorbing barrier (perhaps it would be more correct, in this problem, to say it is the *fly* that is absorbed.)

Here's your problem: What is the distribution of the number of steps it takes the spider to reach the fly, and what is the average number of steps? I feel confident in saying that a theoretical solution for the web of Figure 19.1 isn't going to be easy; therefore, write a computer simulation. In doing that, however, be as general as possible; i.e., create a data structure mechanism that allows you to describe an arbitrary web, not just the one in Figure 19.1. As always with a simulation, it is good practice to at least partially validate your program and, with the ability to input an arbitrary web, you can do this. That is, to test your program, input a one-dimensional web of length $n$, with the spider at $x = 0$ (vertex 1) and the fly at $x = n$ (vertex 2). If your program is correct, it should produce values reasonably close to $n^2$ for the average number of steps taken by the spider to reach the fly.

The random walk you are to consider in this assignment is a very structured random process, constrained by the web. Historically, however, one of the earliest and most famous random walks was the very unstructured one of Brownian motion. Observation under a microscope of small particles suspended in water shows that they move in a jerky, random way, a motion caused by the haphazard bombardment of the particles by water molecules. Indeed, Brownian motion is powerful indirect experimental evidence of the physical reality of molecules (and, hence, of atoms). This motion was first reported by the Scottish botanist Robert Brown (1773–1858) in 1828 (hence its name), and first treated mathematically by Albert Einstein (1879–1955) in 1905. We can easily simulate Brownian motion on a computer as is done by the MATLAB program

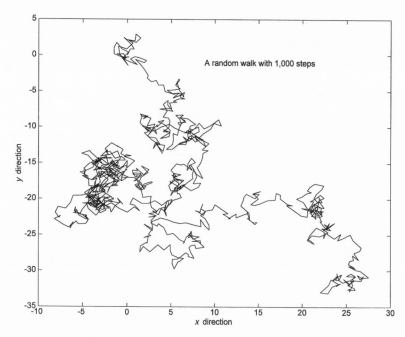

A random walk with 1,000 steps

Figure 19.2 Two-Dimensional Brownian Motion

**brownian.m** (program 5). A typical result is shown in Figure 19.2. The actual motion, of course, is three-dimensional, but this simulation is for the simpler case of two dimensions.

## 20. Reliably Unreliable

Under realistic physical assumptions, it can be shown theoretically (and confirmed by actual observation of real hardware) that many electronic systems fail according to the following probabilistic rule: The probability the system fails sometime during the time interval $x = 0$ (when the system is "turned on") to $x = T$ is $P_F(0 \leq x \leq T) = 1 - e^{-aT}$

$= P_F(T)$, where $a$ is a positive constant characteristic of the system (see more about $a$ below).

This rule has two immediate special values, both of which make obvious physical sense: $P_F(0) = 0$ and $P_F(\infty) = 1$. The first expression says the system cannot fail until it is turned on, and the second expression says the system is certain to fail if we wait long enough.

Now, let's write $X$ as the random variable that denotes the lifetime of the system; i.e., $X$ takes on values that are the lifetimes of identically constructed copies of the system. If $f_X(x)$ is the probability density function of $X$, then we interpret the product $f_X(x)\Delta x$ as the probability that the system has a lifetime somewhere between $x$ and $x + \Delta x$. We can, therefore, write

$$f_X(x)\Delta x = \begin{cases} \text{probability the system fails} \\ \text{sometime between 0 and } x + \Delta x \end{cases}$$
$$- \begin{cases} \text{probability the system fails} \\ \text{sometime between 0 and } x. \end{cases}$$

With the above as background, here is your four-part assignment.

1. Using the expression for $P_F(T)$, show that, as $\Delta x \to 0$, the pdf for $X$ is exponential; i.e., show that

$$f_X(x) = a\, e^{-ax}, x \geq 0$$
$$= 0, x < 0 .$$

2. Calculate the average lifetime of the system, $E(X) = \int_0^\infty x f_X(x)\, dx$, and thus arrive at a physical interpretation for what the constant $a$ represents.

Suppose next that our electronic system is, in fact, the flight control computer for the next generation of commercial jet airliners. A failure of this computer during a flight would be catastrophic and so, to enhance safety, three of these computers are connected in parallel. That is, each computer receives the same instantaneous input signals and, if working properly, each will compute the same instantaneous output control signals. Let's call the parallel connection of the three systems the SYSTEM and assume that the systems are independent.

All three output signal streams are fed into a special circuit that compares them with each other. If all three agree, all is judged well, and if any two agree (while the third is the "odd system out"), then the odd system is judged to have failed in some manner and is switched off. (The special comparator circuit is often called a "majority vote-taker.") The SYSTEM thus continues to work even after one of the parallel connected systems has failed; indeed, the SYSTEM continues to work until the second system failure occurs, an event detected by the comparator when the two initially surviving systems begin to disagree.

The parallel redundancy of the SYSTEM is intended to provide improved reliability. To see if it really does (and if so, by how much), let $Z$ be the random variable denoting the lifetime of the SYSTEM. We'll write $F_Z(z)$ as the distribution function of $Z$, i.e.,

$$F_Z(z) = P(Z \leq z) = \text{probability the SYSTEM fails by time } z.$$

3. Express $F_Z(z)$ in terms of $f_X(x)$ and $F_X(x)$, where, of course, $F_X(x) = \int_0^x f_X(u) \, du$. In the same manner, the pdf of the SYSTEM can be found by writing $F_Z(z) = \int_0^z f_Z(u) \, du$, and so $f_Z(z) = \frac{dF_Z(z)}{dz}$.

4. Use the $f_Z(z)$ from part 3 to find the average value of the lifetime of the SYSTEM, i.e., calculate

$E(Z) = \int_0^\infty z f_z(z)\, dz$. *Compare $E(Z)$ with $E(X)$*; are you surprised by the fact that $E(Z) < E(X)$? If you aren't, why not? After all, isn't that a paradox? Why build an expensive, redundant SYSTEM if, on average, it will fail *sooner* than does an individual system, on average? But, of course, engineers do build such redundant SYSTEMs; can you explain why? There *is* a probabilistic reason.

Finally, here is an interesting military illustration of vote-taking. When a ground-based radar missile site in a war zone detects an incoming aircraft, it challenges the target by sending a friend/foe query via a standard radio channel. If the correct friend/foe "code of the day" is received in reply, then the aircraft is not fired on. Suppose that the channel fails at random, 10 percent of the time, to work properly. Then, 10 percent of friendly aircraft will be attacked. Now, suppose that the missile site issues its challenge over three independent channels, and will not fire if a proper reply is received over at least two of the channels. The probability that a friendly aircraft is properly identified is then $\left(\frac{9}{10}\right)^3 + 3\left(\frac{9}{10}\right)^2 \left(\frac{1}{10}\right) = 0.972$, which means that a friendly aircraft will be attacked only 2.8 percent of the time. This is a reduction, by a factor of almost four, of the erroneous attacks.

## 21. When Theory Fails, There Is Always the Computer

All the problems in this book, so far (with the exception of the first one), have depended on rather sophisticated mathematical analyses and/or computer simulations for their solutions. To complete the circle of difficulty with the final problem, I will end the book with a problem that again uses only the most elementary of arithmetic methods (addition and subtraction). And yet, the actual performing of the required calculations, by hand, is simply too horrible to even think of doing. With MATLAB, however, it will all be duck soup. And best of all, it's actually a practical problem with a multitude of real-life applications. So, with that big buildup, here it is.

Any real construction project of any complexity, such as building a house, dam, airplane, bridge, or computer network of hardware and software is a vast tapestry of interconnected tasks; perhaps hundreds or even thousands of tasks. Some of these tasks can be done independently of other tasks, but they also may depend on yet other tasks (e.g., when building a house you can connect the electrical service entrance to the power utility line either before or after you start the bathroom plumbing, but you have to put the electrical wires through the wall studs before you put up the drywall).

It is very important for the people in charge of any large project to know how long the project will take under the assumption that each task is finished on time, and also to identify those tasks that are critical—i.e., the particular tasks that will delay the completion of the project if they take longer to finish than originally thought. Not all the tasks in most complex projects are critical; they have what is called *slack* (or sometimes *float*) *time*, which is the time delay they can suffer in getting done and yet still not affect the ultimate project completion date. A task with zero slack time (a critical task) is also called a *bottleneck* (or *choke*) task; if it is delayed then the whole project is delayed.

One of the impressive achievements in the analysis and control of complex projects, which came out of the 1950s, is called PERT (Program Evaluation and Review Technique) or, alternatively, CPM (Critical Path Method). The major distinction between the two is that CPM is deterministic (I'll discuss it in detail here), while PERT has a probabilistic flavor (and will form your assignment). CPM grew out of a joint effort between Remington Rand Univac and the E. I. duPont de Nemours Company, while PERT was developed under U.S. Navy sponsorship for use in the Polaris Weapons Systems Program. Indeed, the use of PERT was mandatory for contractors in all U.S. government weapons procurement programs in the 1960s, including even the early U.S. space program (as mentioned in the caption to the frontispiece photograph of the author in 1964). Today, PERT-like software is readily available for use on home computers, e.g., the Microsoft application called *Project*.

To explain the CPM algorithm, I'll use Figure 21.1 as an illustration. It shows a project displayed as an upward-flowing graph of tasks. The arrows show how the tasks are interrelated; each task has one or more *immediate predecessor* tasks (i.e., tasks that flow directly into it), and one or more *immediate successor* tasks (i.e., tasks into which it directly flows). For example, in Figure 21.1, the immediate predecessor tasks of task #5 are tasks #2 and #3, and the immediate successor tasks of task #5 is the single task #7. The interior of each task box contains a single number that denotes the time to complete that task once it has been started (in some convenient, common unit of time).

It will turn out to be convenient in the CPM algorithm to have every project graph begin with what is called a *pseudo-task*; i.e., task #1 is always a nonreal task that takes zero time to perform. So, if a project has six real tasks, as

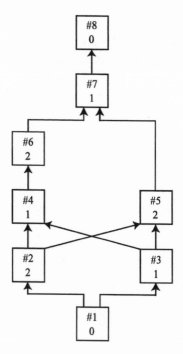

Figure 21.1 Basic CPM Chart

in Figure 21.1, then the first real task is task #2 and the last real task is task #7. It will also prove useful to have every project terminate in a pseudo-task that again takes zero time to do (task #8 in Figure 21.1).

Now, to answer the first of our two fundamental questions—"How long will the project take to do?"—we apply an astonishingly simple rule, one so simple and obvious that at first most students are embarrassed that they didn't think of it themselves: *No task can start until all its immediate predecessor tasks are finished.*

This rule allows the calculation of the earliest start and finish times for each task. So, beginning with task #1, let's start it at time 0 and add to that its completion time (the number inside the task box, which in this case is 0) to get its finish time $(0 + 0 = 0)$. Let's agree to write the earliest start time of any task at the left of the task box and the earliest finish time at the right of the box. Next, follow the arrows from task #1 to all its immediate successors (tasks #2 and #3); they all start when task #1 is done, i.e., at time 0, which we write at the left of each immediate successor task box. Then, we add their completion times to their earliest start times to get their earliest finish times, which are then written at the right of the task boxes.

Next, follow the arrows out of each of the above task boxes to their immediate successors and repeat this process (but don't forget, no task can start until *all* its immediate predecessors are finished; i.e., a task starts at the maximum of the finish times of all of its immediate predecessor tasks) until you reach the final pseudo-task. The result is shown in Figure 21.2, which shows that the project of Figure 21.1 requires 6 units of time to complete.

What are the critical tasks for our project? To answer this, the second of our two fundamental questions, we now apply a new (but still simple) rule: *A task must be finished before any of its immediate successor tasks can start.*

This rule allows the calculation of the latest start time for each task. Here's how.

Starting at the top of the project graph (i.e., with task #8), subtract the completion time from the finish time to get that task's latest start time $(6 - 0 = 6)$. If it starts later than this time, then the project will take longer than 6 time units to do. This tells us that task #8 must start no later than at time $= 6$ and so, according to our new rule, all its immediate predecessor tasks must be finished by that time. In this case, that means that task #7 must be done by time $= 6$, which I've written at the right of the task box (in a circle, to distinguish it from the earliest finish time) in

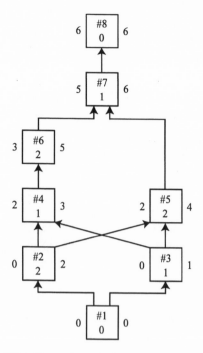

Figure 21.2 CPM Chart Showing Maximum Taks-Finish Times

Figure 21.3. Subtracting the completion time of task #7 from its latest finish time ($6 - 1 = 5$) gives the latest start time for task #7 (which I've written at the left of the task box, again in a circle to distinguish it from the earliest start time). This process is continued all the way back to task #1, with the result shown in Figure 21.3.

Notice that some of the real task boxes have equal earliest and latest start times (tasks #2, 4, 6, and 7), while all the rest of the real task boxes have latest start times greater than their earliest start times. The former are tasks with zero slack time (i.e., bottleneck tasks), while the latter, with positive slack time, can suffer a maximum delay in

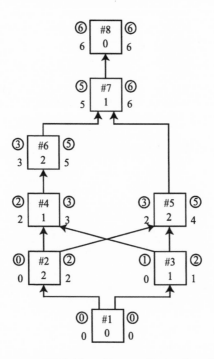

Figure 21.3   CPM Chart Showing Latest Start Times

starting equal to the difference between the two start times without delaying the completion of the project.

This algorithm is easy to understand, mind-deadening to actually follow (imagine a project with thousands of interconnected task boxes), and very easy to program. Indeed, the MATLAB program **cpm.m** (program 6) does it, where the format of entering the task graph for a project is explained in the comment fields of the data matrix statements. The particular entries shown in **cpm.m** are for the project of Figure 21.1, and when **cpm.m** is executed it produces an output of:

Total time required for project = 6

| Task | Early start time | Slack |
|------|------------------|-------|
| 2 | 0 | 0 |
| 3 | 0 | 1 |
| 4 | 2 | 0 |
| 5 | 2 | 1 |
| 6 | 3 | 0 |
| 7 | 5 | 0 |

Here's your assignment: First, analyze the project graph of Figure 21.4 using the CPM algorithm (Project A), then verify your solution by entering the appropriate data statements in **cpm.m** and running it. Next, modify **cpm.m** to incorporate the PERT feature of a *variable* completion time. That is, instead of a single number inside a task box, let there be two numbers and assume the completion time is uniformly distributed between them; e.g., if, in Figure 21.4, the box for task #6 contained the numbers 8 and 12 (rather than the actual single number of 10), then that task could be completed in as little as 8 time units or as long as 12 time units.

Indeed, suppose Figure 21.4 is replaced with Figure 21.5 (Project B). Make the PERT-modified **cpm.m** run through the solution 1,000 times, each time randomly selecting a completion time for each task and also keeping track of the bottleneck tasks for each solution. Note that, because the randomly selected completion times are used as address pointers into a matrix, they must be integers; e.g., if the minimum and maximum completion times for a task are 8 and 12, then your modified program should, for each of the 1,000 solutions, pick uniformly from the integers 8, 9, 10, 11, and 12.

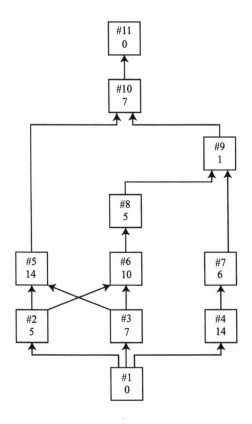

Figure 21.4   CPM/PERT Project A

After the 1,000 solutions have been computed, your new program is to print a histogram of the project completion time, as well as estimates of the probability that each task is a bottleneck task. This is a far more realistic and useful solution than the one produced by the original deterministic CPM algorithm.

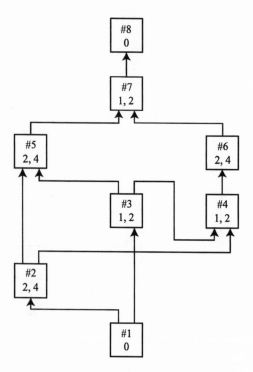

Figure 21.5    CPM/PERT Project B

The Solutions

## 1. How to Ask an Embarrassing Question

Suppose there are $m$ people in your survey and that we let $Y$ denote the total number of YES answers. With probability $\frac{1}{2}$, the coin shows tails on the first flip, so $\frac{1}{2}m$ people will answer the EQ. (This argument is implicitly assuming that $m$ is large.) For the other $\frac{1}{2}m$ people, who flip the coin a second time, half will answer YES (the second flip showed heads) and half will answer NO (the second flip did not show heads). That is, $\frac{1}{4}m$ YES answers are for the non-EQ. So, $Y - \frac{1}{4}m$ YES answers are for the EQ, generated from $\frac{1}{2}m$ people. Thus, the answer to your question is simply

$$\frac{Y - \frac{1}{4}m}{\frac{1}{2}m} = 2\left(\frac{Y}{m} - \frac{1}{4}\right).$$

For $m = 10{,}000$ and $Y = 6{,}230$ the result is

$$2\,(0.623 - 0.25) = 0.746$$

for the fraction of the people in your survey who practice the private act. This is probably a good answer, too, as nobody's privacy was violated and so there was no reason to lie.

This clever idea is relatively new; you can read its original presentation in the paper by Stanley L. Warner, "Randomized Response: A Survey Technique for Eliminating Evasive Answer Bias," *Journal of the American Statistical Association*, vol. 60, March 1965, pp. 63–69.

## 2. When Idiots Duel

You can solve this problem exactly as with the original duel. Therefore, let's write out all the sample points on which A wins and then add the probabilities for all of the points to get the total probability of A's winning. As before, define the two events:

$$a = \{\text{gun fires for A}\}$$
$$x = \{\text{gun does not fire}\}.$$

The sample points on which A wins are

| | | | | |
|---|---|---|---|---|
| *a* | | | | |
| *x* | *xx* | *a* | | |
| *x* | *xx* | *xa* | | |
| *x* | *xx* | *xxa* | | |
| *x* | *xx* | *xxx* | *xxxx* | *a* |
| *x* | *xx* | *xxx* | *xxxx* | *xa* |
| *x* | *xx* | *xxx* | *xxxx* | *xxa* |
| *x* | *xx* | *xxx* | *xxxx* | *xxxa* |
| *x* | *xx* | *xxx* | *xxxx* | *xxxxa* |

and so on.

This is probably enough to show you the pattern. Thus,

$$P(A) = \frac{1}{6} + \left[\left(\frac{5}{6}\right)^3 + \left(\frac{5}{6}\right)^4 + \left(\frac{5}{6}\right)^5\right]\frac{1}{6} + \left[\left(\frac{5}{6}\right)^{10} + \left(\frac{5}{6}\right)^{11}\right.$$
$$+ \left(\frac{5}{6}\right)^{12} + \left(\frac{5}{6}\right)^{13} + \left(\frac{5}{6}\right)^{14}\right]\frac{1}{6} + \left[\left(\frac{5}{6}\right)^{21} + \left(\frac{5}{6}\right)^{22}\right.$$
$$+ \left(\frac{5}{6}\right)^{23} + \left(\frac{5}{6}\right)^{24} + \left(\frac{5}{6}\right)^{25} + \left(\frac{5}{6}\right)^{26} + \left(\frac{5}{6}\right)^{27}\right]\frac{1}{6}$$
$$+ \left[\left(\frac{5}{6}\right)^{36} + \ldots + \left(\frac{5}{6}\right)^{44}\right]\frac{1}{6} + \ldots .$$

Or,

$$P(A) = \frac{1}{6}\left[1 + \left(\frac{5}{6}\right)^3\left\{1 + \left(\frac{5}{6}\right) + \left(\frac{5}{6}\right)^2\right\} + \left(\frac{5}{6}\right)^{10}\left\{1\right.\right.$$
$$+ \left(\frac{5}{6}\right) + \left(\frac{5}{6}\right)^2 + \left(\frac{5}{6}\right)^3 + \left(\frac{5}{6}\right)^4\right\} + \left(\frac{5}{6}\right)^{21}\left\{1 + \left(\frac{5}{6}\right)\right.$$
$$+ \left(\frac{5}{6}\right)^2 + \left(\frac{5}{6}\right)^3 + \left(\frac{5}{6}\right)^4 + \left(\frac{5}{6}\right)^5 + \left(\frac{5}{6}\right)^6\right\}$$
$$+ \left(\frac{5}{6}\right)^{36}\left\{1 + \left(\frac{5}{6}\right) + \left(\frac{5}{6}\right)^2 + \left(\frac{5}{6}\right)^3 + \left(\frac{5}{6}\right)^4\right.$$
$$+ \left(\frac{5}{6}\right)^5 + \left(\frac{5}{6}\right)^6 + \left(\frac{5}{6}\right)^7 + \left(\frac{5}{6}\right)^8\right\}$$
$$\left. + \left(\frac{5}{6}\right)^{55}\{1 + \ldots\} + \ldots\right].$$

The accurate numerical evaluation of this expression is an interesting problem in its own right. One could, of course, just punch the above into a calculator, but that gets tiresome rather quickly and, if your finger slips just once, you must start over. Even if you don't make a mistake in data entry, you will get $P(A)$ with only a relatively small number of correct digits.

In fact, a couple of minutes of boring key-pushing on my cheap drugstore calculator gave the result $P(A) \approx 0.524$, which is only just a little bit less than the value for the original duel. That is, giving B two trigger pulls on his first possession of the gun is still not enough to make the duel fair, i.e., to make $P(A) = \frac{1}{2}$. Having the first shot is really important!

Finally, the average duration of the duel under the new rules is still six trigger pulls. From the gun's "point of view," there is simply a finger pulling its trigger. It matters not a bit to the gun if it is A's finger or B's finger (although it matters a great deal, of course, to A and B).

*Postscript*: if the probability analysis done so far is all you're interested in, then you can skip what follows. But, as I mentioned earlier, the precise calculation of $P(A)$ has some interesting challenges all its own. Suppose, for some reason, we aren't happy with the above three-digit approximation, but instead want twenty or thirty (or more) digits. What do we do then? How can a human possibly correctly add enough terms in the series for $P(A)$, as well as avoid the problem of round-off errors, to arrive at a result with such astounding precision?

The answer is, as you may have guessed, to use the computer and MATLAB. MATLAB, in particular, with its symbolic manipulation capability, provides super-high precision at a price that is, metaphorically, cheaper than dirt. So, first we need to recognize a programmable pattern in the expression for $P(A)$. If you stare at that expression for about five seconds, you should spot the following pattern: $P(A)$ is one-sixth of what is contained in the brackets, which itself is one plus terms that obey the following basic rule—$\left(\frac{5}{6}\right)$ raised to a power, times a finite-length geometric sum.

There are two specific rules we need to discover now: the rule for the value of the power in each term, and the rule for the length of the finite sum in each term.

**Power rule:** *The power value starts at 3, then increases to 10, then increases to 21, then increases to 36, then increases to 55, and so on. This can be restated as follows: The power starts at 3, then we add 7, then we add 11, then we add 15, and so on. The amount we add to get the next power is just the amount we added the previous time plus 4.*

**Sum length rule:** *The first sum is $1 + \left(\frac{5}{6}\right) + \left(\frac{5}{6}\right)^2$. The next sum adds two more terms, as does the next sum, and so on.*

These rules are easy to program (see **idiots2.m**, Program 7) and, combined with symbolic manipulation to avoid round-off error, we'll be just about done. When doing "normal" arithmetic, MATLAB replaces $\frac{5}{6}$ with $0.83333\ldots$ out to 32 digits and then simply stops. This leads to errors that accumulate at every step. In symbolic form, however, MATLAB keeps $\frac{5}{6}$ as $\frac{5}{6}$; e.g., if you ask MATLAB to compute $x + y$ where $x = \text{sym}$ ("1/7") and $y = \text{sym}$ ("1/13") it will return the *exact* answer of 20/91. There is no error accumulation. So, the end result, $P(A)$, with 25 correct digits, is

$$P(A) = 0.5239191275550995247919843,$$

a number I feel safe in saying has never appeared in print before until now.

### 3. Will the Light Bulb Glow?

This is actually an easy problem for which it is also, however, quite easy to lose your way. The trick is to resist the temptation to blast through the analysis too quickly. I will take it step by step, slow and easy. Since both questions involve sheets of switches, let's start by determining the probability that there is an electrical path from one end of a sheet to the other. This is the sort of problem in which it is easier to calculate the probability that there is *not* such a path, and then to subtract that probability from 1 to get the probability that there *is* such a path. After all, there either is or isn't a path; it's one or the other. This is a powerful trick to remember.

So, first consider a row of switches. For a row to form a path, all $n$ switches must be closed; this occurs with probability $p^n$. Thus, a row does *not* form a path with probability $1 - p^n$. For a sheet *not* to contain a path, all $n$ rows must *not* be a path; this occurs with probability $(1 - p^n)^n$. Thus, the probability that a sheet does have at least one path from end to end is $1 - (1 - p^n)^n$. So, finally, if there are $n$ sheets in series, then all $n$ sheets must have at least one path for the bulb to glow, and so the answer for series sheets is

$$P_{series} (n, p) = \left[ 1 - (1 - p^n)^n \right]^n.$$

With the sheets of switches in parallel, none of them has a path from end to end (and so the bulb does not glow) with probability

$$\left[ (1 - p^n)^n \right]^n = (1 - p^n)^{n^2}.$$

Thus, the bulb *does* glow (with parallel sheets) with probability

$$P_{parallel} (n, p) = 1 - (1 - p^n)^{n^2}.$$

The MATLAB program **bulb.m** (Program 8) plots these two expressions for $0 \le p \le 1$ for the two cases of $n = 5$ and $n = 10$, i.e., for the cases of 125 and 1,000 switches, respectively. These two plots, Figures 3.3 and 3.4, tell us three things about our random switch network:

1. The sheets-in-parallel connection, for given values of $n$ and $p$, has a higher probability of the bulb's glowing than does the sheets-in-series connection.

2. For both connections, the probability of the bulb's glowing, for a given $p$, decreases as $n$ increases, i.e., the curves shift to the right as $n$ increases.

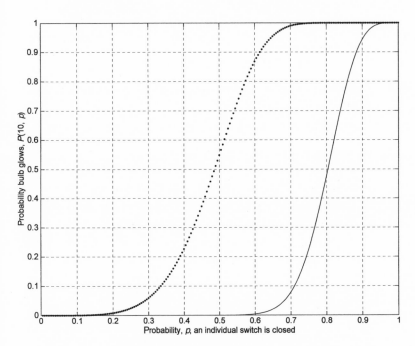

Figure 3.3 Probability of Bulb Glowing, $n = 5$ (solid line for series sheets, dots for parallel sheets)

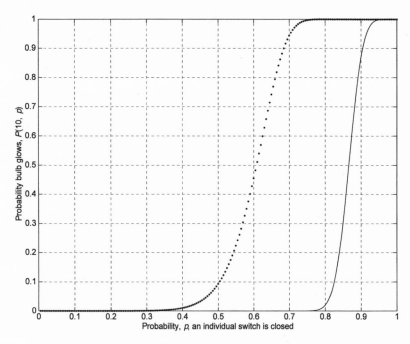

*Figure 3.4*  Probability of Bulb Glowing, $n = 10$ (solid line for series sheets, dots for parallel sheets)

3. For both connections, there is a "threshold effect" in $p$, i.e., for small $p$, both connections have a small probability of conducting (bulb glowing), but as $p$ increases toward a critical value (which depends on $n$), both connections flip over quickly to be "very probably" conducting. For $n = 10$, the threshold for parallel sheets is about $p = 0.6$, and for series sheets it is about $p = 0.85$.

Finally, for the Markov chain problem we have, since $p(t + 1) = p(t)P$,

$$p(1) = p(0)\ P$$
$$p(2) = p(1)\ P = p(0)\ P^2$$
$$p(3) = p(2)\ P = p(0)\ P^3$$

.

.

.

$$p(t) = p(0) \, P^t.$$

So, all we need to do, to find $p(t)$, is to multiply the initial state row vector $p(0)$ with the $t$ th power of the state transition matrix. The answer to your question is just $p_3(t)$, the rightmost element in the state vector, but in fact, why not plot *all* the elements, since they are there, too, for free? The program **markov.m** (Program 9) performs all of the required calculations, and Figure 3.5 shows the results. $p_0(t)$ monotonically drops toward zero, $p_1(t)$ and $p_2(t)$ first rise and then drop toward zero with $p_2(t)$ lagging $p_1(t)$, and $p_3(t)$ rises monotonically toward unity. This all makes physical

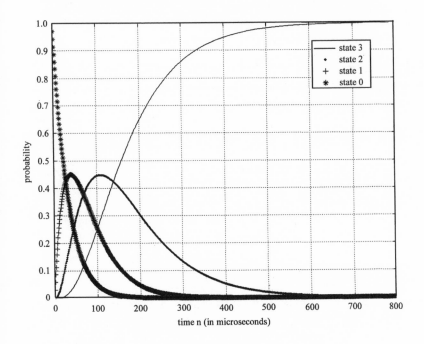

Figure 3.5 Probability the Path Is in State $k$ at Time $n$

sense, as the state of the path evolves strictly in one direction, from 0 to 3. The sum $\sum_{k=0}^{3} p_k(t) = 1$ for all $t$, since the path is always in some state. The curve for $p_3(t)$ shows, further, that $p_3(t) = \frac{1}{2}$ when $t =$ about 150 microseconds; i.e., this is the median time it takes for the bulb to start glowing. Finally, the rows of $P$ all sum to one, individually, because $P(i, j)$ is the probability of the path moving from state $i$ to state $j$ in a single time increment. Since the path in state $i$ at time $t$ has to be in *some* state at time $t + 1$, then

$$\sum_j P(i, j) = 1$$

for each $i$, i.e., for each row of $P$.

As a last note, we can directly derive the so-called steady-state or end-state probabilities with the following argument. Let's write $\pi_i = \lim_{t \to \infty} p_i(t)$. Then, since by definition the steady-state probabilities don't change from $t$ to $t + 1$, we have

$$[\pi_0\ \pi_1\ \pi_2\ \pi_3] = [\pi_0\ \pi_1\ \pi_2\ \pi_3] \begin{bmatrix} 0.97 & 0.03 & 0 & 0 \\ 0 & 0.98 & 0.02 & 0 \\ 0 & 0 & 0.99 & .01 \\ 0 & 0 & 0 & 1 \end{bmatrix}.$$

Or, writing out the matrix multiplication explicitly, we get the four equations

(a) $\pi_0 = 0.97\ \pi_0$
(b) $\pi_1 = 0.03\ \pi_0 + 0.98\ \pi_1$
(c) $\pi_2 = 0.02\ \pi_1 + 0.99\ \pi_2$
(d) $\pi_3 = 0.01\ \pi_2 + \pi_3$.

Also, as mentioned above the path has to be in some state, at each instant, and so we have the fifth equation,

(e) $\pi_0 + \pi_1 + \pi_2 + \pi_3 = 1$.

Equations (a), (b), and (c) tell us that $\pi_0$, $\pi_1$, and $\pi_2$ must each be zero. Equation (d) then tells us nothing, but Equation (e) says $\pi_3 = 1$.

## 4. The Underdog and the World Series

It may seem silly to actually say this, but the key observation to solving this problem is: The team that wins the Series wins the last game played. That is, the number of different ways the series can be played in five games (for example) is not the $\binom{7}{5} = \frac{7!}{5!\,2!} = 21$ ways that many beginning students write, but rather is $\binom{4}{3} = \frac{4!}{3!\,1!} = 4$ ways. This is because if the series lasts five games, then there simply aren't seven games from which to select the four that the series-winning team wins. Rather, to win in five games, the series-winning team wins the fifth (last) game as its fourth victory, and so there are $\binom{4}{3} = 4$ ways for it to have won its other three victories from the earlier four games. Thus, if we denote the number of ways the series lasts 4, 5, 6, and 7 games by $N_4$, $N_5$, $N_6$, and $N_7$, respectively, we have

$$N_4 = \binom{3}{3} = 1 \qquad N_5 = \binom{4}{3} = 4$$

$$N_6 = \binom{5}{3} = 10 \qquad N_7 = \binom{6}{3} = 20.$$

With these numbers in hand, we can now immediately write the answer to our question—remember, it's the *weaker* team, with probability $1 - p$ of winning a game, that wins the series in this calculation:

$$\begin{aligned}
P(p) &= N_4(1-p)^4 + N_5(1-p)^4 p + N_6(1-p)^4 p^2 \\
&\quad + N_7(1-p)^4 p^3 = (1-p)^4 + 4(1-p)^4 p \\
&\quad + 10(1-p)^4 p^2 + 20(1-p)^4 p^3.
\end{aligned}$$

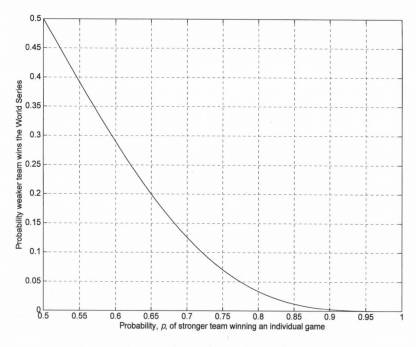

*Figure 4.1* The Stronger Team Does *Not* Always Win

We could multiply all this out, combine terms, and try to simplify, but why bother? It's easier, and more informative, to just plot $P(p)$ versus $p$ on a computer and then read the numbers right off the curve. I've done that (see **underdog1.m** [Program 10] and Figure 4.1), and the numbers can be surprising. For example, if $p = 0.67$ (which means that the stronger team, on average, wins twice as many games as does the weaker team), then the weaker team still has a nearly 17 percent chance of winning the series. The moral is clear: "It ain't over 'til it's over!" Or as an old saying puts it, "It isn't always the cream that floats to the top."

From the historical data, the average duration of the World Series has been, under modern rules,

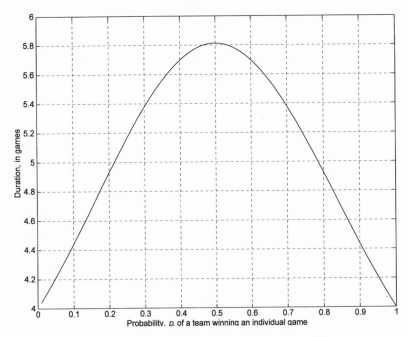

Duration, in games

Probability, $p$, of a team winning an individual game

Figure 4.2    Expected Duration of the World Series

$$\frac{(4 \times 17) + (5 \times 20) + (6 \times 21) + (7 \times 33)}{91} = 5.8 \text{ games.}$$

What does probability theory tell us? The expected or average value of a discrete random variable—as is the duration of the series, which is equal to one of the integers in the finite set (4, 5, 6, 7)—is the sum

$$D(p) = \sum_{k=4}^{7} k P_k(p)$$

where $P_k(p)$ is the probability the series is of duration $k$ games. We already have these probabilities from the first part of this analysis, of course; e.g., $P_5 = N_5 [(1 - p)^4 p + p^4(1 - p)] = 4 [(1 - p)^4 p + p^4(1 - p)]$. Note that $P_5$ accounts for the fact that either team could be the winner of a five-game series. So, we immediately have for our answer the rather cumbersome-looking expression (which I've programmed into **underdog2.m** [Program 11]):

$$D(p) = 4[(1-p)^4 + p^4] + 20\,[(1-p)^4 p + p^4(1-p)]$$
$$+ 60\,[(1-p)^4 p^2 + p^4(1-p)^2] + 140\,[(1-p)^4 p^3$$
$$+ p^4(1-p)^3].$$

Figure 4.2 shows how $D(p)$ varies over the interval $0 \le p \le 1$. We see that, as you'd expect, $D(0) = D(1) = 4$, and that $D\left(\frac{1}{2}\right)$ is the maximum of $D(p)$; thus is 5.8, which is also just the historical average. So, on average, the World Series teams *seem* to have been evenly matched.

## 5. The Curious Case of the Snowy Birthdays

The MATLAB program **esim.m** (Program 11) carries out the dart tossing experiment described in the original problem statement. And, just as the theory developed there predicts, the results are somewhat suggestive of *e*:

| N | Estimates of E (from three simulations) | | |
|---|---|---|---|
| 100 | 2.5641 | 3.1250 | 2.5000 |
| 1,000 | 2.8571 | 2.5907 | 2.6667 |
| 5,000 | 2.6596 | 2.7518 | 2.8233 |

The MATLAB program **randomsum.m** (Program 13) carries out the "add random numbers until the sum reaches at least 1" experiment 100,000 times. And, just as I claimed in the problem statement, if $N$ is the number of numbers required for a sum to reach at least one, then $E(N)$ is fairly close to *e*:

| Simulation | E(N) |
|:----------:|:------:|
| 1 | 2.7193 |
| 2 | 2.7194 |
| 3 | 2.7220 |

Now, why does $E(N) = e$? As I suggested in the problem statement, you can answer this with the same approach used for the Daphne Tams's "snowy birthdays" problem. Let's start by writing $S_n = X_1 + X_2 + \ldots + X_n$, where the $X_i$ are independent and uniform from 0 to 1, and $n = N$ is the first value that gives $S_n \geq 1$.

We can write, for any value of $n$,

$$
\begin{aligned}
P(N > n) &= P(S_1 < 1, S_2 < 1, \ldots, S_n < 1) \\
&= P(X_1 < 1, X_1 + X_2 < 1, \ldots, X_1 \\
&\quad + X_2 + \ldots + X_n < 1) \\
&= P(X_1 + X_2 + \ldots + X_n < 1),
\end{aligned}
$$

where the last equality follows since the sums monotonically increase with increasing $n$.

Next, as argued in the snowy birthday problem,

$$
\begin{aligned}
P(N = n) &= P(N > n - 1) - P(N > n) \\
&= P(X_1 + X_2 + \ldots + X_{n-1} < 1) - P(X_1 \\
&\quad + X_2 + \ldots + X_n < 1) \\
&= P(S_{n-1} < 1) - P(S_n < 1).
\end{aligned}
$$

But (and don't forget that $P(N = 1) = 0$), this says

$$
\begin{aligned}
E(N) &= \sum_{n=1}^{\infty} nP(N = n) = \sum_{n=2}^{\infty} n\,[P(S_{n-1} < 1) - P(S_n < 1)] \\
&= 2\,[P(S_1 < 1) - P(S_2 < 1)] + 3\,[P(S_2 < 1) \\
&\quad - P(S_3 < 1)] \\
&\quad + 4\,[P(S_3 < 1) - P(S_4 < 1)] + \ldots.
\end{aligned}
$$

This infinite summation partially "telescopes" (i.e., adjacent terms partially cancel) to give

$$E(N) = 2P(S_1 < 1) + P(S_2 < 1) + P(S_3 < 1) + \ldots.$$

Since we know $P(S_1 < 1) = 1$, we thus have

$$E(N) = 2 + P(S_2 < 1) + P(S_3 < 1) + \ldots.$$

The probabilities on the right-hand side are, in principle, easy to write down. For example,

$$P(S_2 < 1) = P(X_1 + X_2 < 1) = P(X_2 < 1 - X_1)$$

$$= \int_0^1 \int_0^{1-x_1} dx_2 \, dx_1 = \int_0^1 (1 - x_1) \, dx_1$$

$$= \left. \left( x_1 - \frac{1}{2} x_1^2 \right) \right|_0^1 = 1 - \frac{1}{2} = \frac{1}{2}.$$

Similarly,

$$P(S_3 < 1) = P(X_1 + X_2 + X_3 < 1)$$

$$= \int_0^1 \int_0^{1-x_1} \int_0^{1-x_1-x_2} dx_3 \, dx_2 \, dx_1.$$

Indeed, the general term in sum for $E(N)$, $P(S_n < 1)$, is simply the multiple integral

$$P(S_n < 1) = \int_0^1 \int_0^{1-x_1} \int_0^{1-x_1-x_2} $$
$$ \ldots \int_0^{1-x_1-x_2\ldots-x_{n-1}} dx_n \, dx_{n-1} \ldots dx_2 \, dx_1.$$

Mathematicians somewhat forbiddingly call this integral the "$n$-dimensional hypervolume" bounded by the positive $x_1, x_2, \ldots, x_n$ axes, and the hyperplane $x_1 + x_2 + \ldots + x_n = 1$. If you grind through the integral for the $n = 3$

and $n = 4$ cases (and make no mistakes; they do get progressively more awful) you'll get $P(S_3 < 1) = \frac{1}{6}$ and $P(S_4 < 1) = \frac{1}{24}$.

Staring at these partial results for a while may suggest to you that we can write $P(S_2 < 1) = \frac{1}{2!}$, $P(S_3 < 1) = \frac{1}{3!}$, and $P(S_4 < 1) = \frac{1}{4!}$. And that may further suggest to you that $P(S_n < 1) = \frac{1}{n!}$. In fact, using mathematical induction, one can show that this is indeed the case (but I'm not going to do that here). We have, therefore,

$$E(N) = 2 + \frac{1}{2!} + \frac{1}{3!} + \frac{1}{4!} + \ldots + \frac{1}{n!} + \ldots,$$

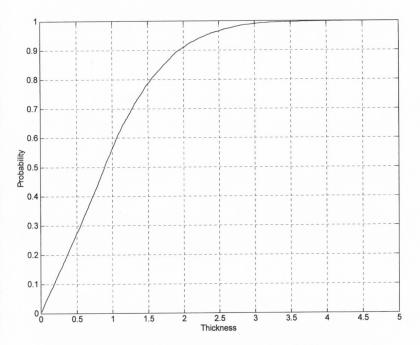

Figure 5.2  Distribution of Total Thickness from 5,000 cuttings, 10 Slices per Cutting

which is the sum for $e$ given in the original statement of the problem, i.e., $E(N) = e$ just as suggested by the numbers produced by **randomsum.m**.

Finally, the MATLAB program **onion.m** (Program 14) simulates the slice-and-dice operations on the unit interval a total of 5,000 times, each time making ten slices. As Figure 5.2 shows, $P(S \le s)$ does appear to increase in a linear fashion, and when I looked at the value of dist(100)/5000 $= P(S \le 1)$, the value was 0.5634, which compares well with the theoretical result of $e^{-\gamma} \approx e^{-0.577} = 0.5616$.

## 6. When Human Flesh Begins to Fail

The probability that a fair coin comes up heads $k$ times in $n$ flips is $\binom{n}{k}\left(\frac{1}{2}\right)^n$. If $N$ people do this independently, then the probability of that is

$$\left[\binom{n}{k}\left(\frac{1}{2}\right)^n\right]^N.$$

Since $k$ can be any integer from 0 to $n$, then for our final result we have $P(N, n) =$ probability $N$ people all get the same number of heads in $n$ flips as

$$P(N,n) = \frac{1}{2^{nN}}\sum_{k=0}^{n}\binom{n}{k}^N.$$

If $n$ is much bigger than ten or so, then there are a fair number of binomial coefficients to evaluate. Of course, since $\binom{n}{k} = \binom{n}{n-k}$, we really have to do only half of them, but that's a small consolation if $n = 100$ and they all have to be raised to the $N$th power.

For the case of $N = 2$ people, we can write the combinatorial identity

$$\sum_{k=0}^{n} \binom{n}{k}^2 = \binom{2n}{n}$$

to make things a lot neater. Then,

$$P(2, n) = \frac{1}{2^{2n}} \binom{2n}{n} = \frac{1}{2^{2n}} \frac{(2n)!}{(n!)^2},$$

which can be further reduced with Stirling's wonderful asymptotic formula for $n!$, i.e., $n! \sim \sqrt{2\pi n} \; n^n \; e^{-n}$. (This extremely useful formula is named after the English mathematician James Stirling (1692–1770), who published it in his 1730 book *Methodus Differentialis*. Ever since, mathematicians have found the search for an elementary derivation to be a temptation that is impossible to resist; in my opinion, it is still hard to prove.) Once all the cancelations are done, you should be able to show that

$$P(2,n) \approx \frac{1}{\sqrt{\pi n}},$$

a result that "gets better" as $n$ gets larger. In fact, it works remarkably well even for small $n$, as you will see below. Such a neat result for other values of $N$ doesn't seem to exist, however, and a computer solution is the practical one for arbitrary $N$ and $n$.

I have written a MATLAB program (which should be easy to convert into just about any other commonly used language) to evaluate $P(N, n)$. This program, called **match.m** (Program 15), needs all those binomial coefficients, of course, and MATLAB, for some odd reason, has no such function in its otherwise impressive math library. So, I have also written the subroutine **binomial.m** (Program 16, used in a couple of other problems in this book), which **match.m** calls with two arguments, $n$ and $k$. Running them gives us the answers we are after:

| $n$ | Flips | | | |
|---|---|---|---|---|
| $N$ | 10 | 50 | 100 | 150 |
| 2 | 0.1762 | 0.0796 | 0.0563 | 0.0460 |
| people 3 | 0.0355 | 0.0073 | 0.0037 | 0.0024 |
| 4 | 0.0076 | $7.1 \times 10^{-4}$ | $2.5 \times 10^{-4}$ | $1.4 \times 10^{-4}$ |

Just for the fun of it, I tried a few other combinations, too:

$$P(100, 10) = 1.29 \times 10^{-61}$$
$$P(200, 5) = 1.87 \times 10^{-101}$$
$$P(300, 3) = 3.24 \times 10^{-128}.$$

What does that last probability mean? Here's one way to visualize how small it is. The Earth is about four billion years old; since there are about 30 million seconds in a year, the Earth is about $10^{17}$ seconds old. Suppose that God, at the formation of the planet, assigned 300 angels the task of each flipping a fair coin (could there be, I wonder, an unfair coin in heaven?) three times and then comparing their individual counts of heads. If they all get the same number—0, 1, 2, or 3—they are to immediately inform God. If they don't all get the same number of heads, then they are to flip again. And again . . . until they do. If those 300 angels could repeat this process $10^{100}$ times per second (a mathematician would call this a frequency of 1 googolhertz), then they have had time to do it $10^{117}$ times. But with a probability of $10^{-128}$ for a match, it is almost certain that God is still waiting. Indeed, given the long, boring nature of this task, perhaps it is more likely that Satan has his people doing this down in hell than God has his doing it up in heaven!

Finally, if we go back and use the simple formula for the case of $N = 2$ people, we see that, for example,

$$P(2, 10) \approx \frac{1}{\sqrt{10\pi}} = 0.1784,$$

which compares well with the result computed directly from the summation. And for $n = 150$, the agreement is even better:

$$P(2, 150) \approx \frac{1}{\sqrt{150\pi}} = 0.0461,$$

which is very close to the MATLAB-computed value of 0.0460.

## 7. Baseball Again, and Mortal Flesh, Too

The probability of winning at least $np$ games out of $n$ games, denoted by $P_n(p)$, is simply the sum of the probabilities of winning exactly $np$ games, $np + 1$ games, $\dots$, $n$ games. Since the number of games played is always an integer (and $np$ may not always be an integer), we start the sum with $s_n = $ first integer equal to or greater than $np$. Thus,

$$P_n(p) = \sum_{k=s_n}^{n} \binom{n}{k} p^k (1-p)^{n-k}.$$

The problem, then, is to plot the ratio

$$\frac{P_{81}(p)}{P_{162}(p)} = \frac{\displaystyle\sum_{k=s_{81}}^{81} \binom{81}{k} p^k (1-p)^{81-k}}{\displaystyle\sum_{k=s_{162}}^{162} \binom{162}{k} p^k (1-p)^{162-k}}$$

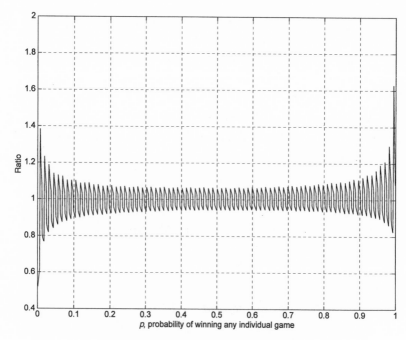

Figure 7.2   Ratio of Probabilities of Winning at Least $np$ Games Out of $n,n = 81/n = 162$, $p$-resolution $= 0.001$

as a function of $p$. This is definitely an expression that only somebody with a lot of time on his or her hands would even dream of evaluating by hand. With a computer, however, it is duck soup. The MATLAB program **baseball.m** (Program 17) performs the calculation, and was used to generate Figure 7.1 in the original problem statement (with a $p$-resolution of 0.01). When run with a $p$-resolution of 0.001, the result is Figure 7.2, which is at least as choppy-looking as is Figure 7.1. Increasing the $p$-resolution (decreasing the step increment between successive $p$-values) will not make the plot smoother, and the reason is because of the discontinuity of the floor command truncation operations when calculating the quantities $s_{81}$ and $s_{162}$.

# 8. Ball Madness

To answer the preliminary question of why the given state transition probabilities for the Ehrenfest ball-exchange process are correct, consider the following arguments.

1. If $k = i + 1$, then the number of black balls in I increases by 1, from $i$ to $i + 1$. This means, first, that before the exchange there are $i$ black balls (and so, $n - i$ white balls) in I, and there are $n - i$ black balls and $i$ white balls in II. So, to increase the number of black balls in I by 1, we must select a white ball from I (with probability $\frac{n-i}{n}$) and a black ball from II (with probability $\frac{n-i}{n}$). Since the two selections are independent, then

$$P(i, k) = \left(\frac{n-i}{n}\right)\left(\frac{n-i}{n}\right) = \left(\frac{n-i}{n}\right)^2.$$

The constraint $i < n$ simply says that there are only $n$ black balls in all, and so the transition from $i = n$ to $i = n + 1$ is not possible.

2. If $k = i$, then the number of black balls in I remains unchanged by the exchange. This can happen in two ways. First, a black ball could be selected from each urn and simply swapped. Or, a white ball could be selected from each urn and swapped. Either way, the net result is no change. So,

$$P(i, k) = \left(\frac{i}{n}\right)\left(\frac{n-i}{n}\right) + \left(\frac{n-i}{n}\right)\left(\frac{i}{n}\right) = \frac{2i(n-i)}{n^2}.$$

3. If $k = i - 1$, then the number of black balls in I decreases by 1, from $i$ to $i - 1$. This means a black

ball is selected from I (with probability $\frac{i}{n}$) and a white ball is selected from II (with probability $\frac{i}{n}$). Thus,

$$P(i, k) = \left(\frac{i}{n}\right)\left(\frac{i}{n}\right) = \left(\frac{i}{n}\right)^2.$$

The constraint $i > 0$ makes the obvious statement that the transition from $i = 0$ to $i = -1$ is impossible.

4. There are no other possible cases, and so

$$P(i, k) = 0 \text{ otherwise.}$$

Now, for the main problem, let's define $N$ as the number of draws before we again draw a previously drawn ball; we want to calculate $E(N)$. What is the probability that $N = 1$, i.e., what is the probability that we have just one draw before we draw a previously drawn ball? One the first draw we draw some ball (any of the $n$). Then we must draw that same ball on the second draw for $N$ to be 1. That clearly happens with probability $\frac{1}{n}$ and, since the two draws are independent, then

$$N = 1 \text{ with probability } \frac{n}{n} \times \frac{1}{n} = \frac{1}{n}.$$

What is the probability that $N = 2$? On the first draw we draw any ball, and on the second draw we draw any of the other remaining $n - 1$ balls. Then, on the third draw, we must draw one of the two previously drawn balls, i.e.,

$$N = 2 \text{ with probability } \frac{n}{n} \times \frac{n-1}{n} \times \frac{2}{n} = \frac{2(n-1)}{n^2}.$$

Now, just one more time. What is the probability that $N = 3$? On the first draw, we draw any ball, and on the second draw we draw any of the remaining $n-1$ balls, and on the third draw we draw any one of the remaining $n-2$ balls. Then, on the fourth draw we must draw one of the three previously drawn balls, and so

$$N = 3 \text{ with probability } \frac{n}{n} \times \frac{n-1}{n} \times \frac{n-2}{n} \times \frac{3}{n} = \frac{3(n-1)(n-2)}{n^3}.$$

You should have the pattern by now. So,

$$E(N) = \sum_{k=1}^{N} kP(N=k) = 1 \times \frac{1}{n} + 2 \times \frac{2(n-1)}{n^2}$$

$$+ 3 \times \frac{3(n-1)(n-2)}{n^3}$$

$$+ \ldots + n \frac{n(n-1)(n-2) \ldots 3 \times 2 \times 1}{n^n}$$

$$= \frac{1}{n} + 2^2 \frac{n-1}{n^2} + 3^2 \frac{(n-1)(n-2)}{n^3} + \ldots$$

$$+ n^2 \frac{(n-1)(n-2) \ldots 3 \times 2 \times 1}{n^n}.$$

The answer to the second equation (call it $T$), for the largest number of drawings for which the probability of not having a repetition remains greater than $\frac{1}{2}$, is the largest integer $T$ for which the following inequality holds:

$$\frac{n}{n} \times \frac{n-1}{n} \times \frac{n-2}{n} \ldots \frac{n-(T-1)}{n} > \frac{1}{2}.$$

This expression, and the one for $E(N)$, are both easy to program; I have done so in the MATLAB program **balls.m**

(Program 18). The results produced, for various numbers of balls ($n$) are:

| $n$ | $E(N)$ | $T$ |
|---|---|---|
| 10 | 3.6602 | 4 |
| 100 | 12.2100 | 12 |
| 1,000 | 39.3032 | 37 |
| 10,000 | 124.9991 | 118 |
| 100,000 | 395.9997 | 372 |

I think these values for both $E(N)$ and $T$ are astonishingly small, in violent contradiction to just about everyone's intuition. They also suggest that both $E(N)$ and $T$ are proportional not to $n$, but to $\sqrt{n}$ (which can, in fact, be shown to be the case using more analysis than I care to do here).

## 9. Who Pays for the Coffee?

We have $N-1$ people, each with a fair coin, and an Nth person with a coin biased such that $P(\text{heads}) = q$ and $P(\text{tails}) = 1 - q$. To get an odd person out on a given simultaneous flipping, $N-1$ of them must get one result and one get the other result. This can happen in the following ways:

1. The $N-1$ people with fair coins all get heads and the person with the biased coin gets tails; the probability of this is $\left(\frac{1}{2}\right)^{N-1}(1-q)$.

2. The $N-1$ people with fair coins all get tails and the person with the biased coin gets heads; the probability of this is $\left(\frac{1}{2}\right)^{N-1} q$.

3. The person with the biased coin gets heads, as do $N-2$ of the $N-1$ people with fair coins, and the

remaining person with a fair coin gets tails. Since there are $N-1$ ways to pick the person with a fair coin who is the odd person out, then the probability of this is $q \left(\frac{1}{2}\right)^{N-2} (N-1) \frac{1}{2}$.

4. The person with the biased coin gets tails, as do $N-2$ of the $N-1$ people with fair coins, and the remaining person with a fair coin gets heads. Since there are $N-1$ ways to pick the person with a fair coin who is the odd person out, then the probability of this is $(1-q) \left(\frac{1}{2}\right)^{N-2} (N-1) \frac{1}{2}$.

So, the total probability of an odd person out is

$$\left(\frac{1}{2}\right)^{N-1} (1-q) + \left(\frac{1}{2}\right)^{N-1} q + q \left(\frac{1}{2}\right)^{N-2} (N-1) \frac{1}{2}$$

$$+ (1-q) \left(\frac{1}{2}\right)^{N-2} (N-1) \frac{1}{2}$$

$$= \left(\frac{1}{2}\right)^{N-1} [(1-q) + q]$$

$$+ \left(\frac{1}{2}\right)^{N-1} (N-1) [q + (1-q)]$$

$$= \left(\frac{1}{2}\right)^{N-1} + \left(\frac{1}{2}\right)^{N-1} (N-1)$$

$$= N \left(\frac{1}{2}\right)^{N-1}.$$

But this is exactly the result found in the original problem statement for the case of all $N$ coins' being fair. The presence of a biased coin has no effect. The average duration of a game, therefore, will also be unchanged. To check this conclusion by a simulation, I have modified **odd.m** to

**biased.m** (Program 19) which asks for the value of $q$. Here's what I got when I ran **biased.m**, results that support the theoretical conclusion.

| $N$ | $2^{N-1}/N$ | $q$ | Average game duration (flips) from **biased.m** (three simulations) | | |
|---|---|---|---|---|---|
| 3 | 1.333 | 0 | 1.336 | 1.348 | 1.337 |
| | | .3 | 1.316 | 1.346 | 1.339 |
| | | .9 | 1.321 | 1.324 | 1.308 |
| 4 | 2 | 0 | 2.064 | 2.02 | 1.95 |
| | | .3 | 2.032 | 2.104 | 1.914 |
| | | .9 | 2.051 | 2.028 | 2.044 |
| 5 | 3.2 | 0 | 3.15 | 3.181 | 3.161 |
| | | .3 | 3.167 | 3.301 | 3.14 |
| | | .9 | 3.242 | 3.302 | 3.411 |
| 6 | 5.333 | 0 | 5.323 | 5.613 | 5.444 |
| | | .3 | 5.21 | 5.187 | 5.294 |
| | | .9 | 5.19 | 5.471 | 5.552 |

## 10. The Chess Champ versus the Gunslinger

As suggested, define $C(k, n)$ to be the probability that the champ is ahead by $k$ points at the end of the $n$th game. For example, $C(0, n)$ is the probability that match is tied at the end of the $n$th game. Now, recalling Laplace's famous words, "We ought to regard the present state of the Universe as the effect of its preceding state and as the cause of its succeeding states," there are just three ways that the champ can be ahead by $k$ points at the end of the $n$th game:

1. The champ was ahead by $k$ points at the end of the $n - 1$st game, and the $n$th game was a draw;

2. The champ was ahead by $k + 1$ points at the end of the $n - 1$st game, and lost the $n$th game;

3. The champ was ahead by $k - 1$ points at the end of the $n - 1$st game, and won the $n$th game.

So, we immediately have the difference equation

$$C(k, n) = C(k, n-1)q + C(k+1, n-1)(1-p-q) \\ + C(k-1, n-1)p.$$

If we can directly calculate some preliminary values of $C(k, n)$, then we can use this recursive equation to calculate those values of $C(k, n)$ that we cannot find directly. For example, $C(k, k)$ is easy to calculate, since there is only one way the champ can be ahead by $k$ points at the end of the $k$th game: by winning all $k$ games. So,

$$C(k, k) = p^k.$$

Another special case we can calculate directly is $C(0, n)$, the probability that the champ and challenger are tied at the end of the $n$th game. Suppose $n$ is odd. The only way to have a tied match is to have had an odd number of draws, thereby leaving an even number of won games split equally between the champ and the challenger. If there are $d$ drawn games ($d = 1, 3, 5, \ldots, n$) then the champ wins $\frac{1}{2}(n - d)$ games and the challenger wins $\frac{1}{2}(n - d)$ games. The probability of any particular such situation is

$$p^{\frac{1}{2}(n-d)} (1-p-q)^{\frac{1}{2}(n-d)} q^d$$

and the number of such situations is

$$\binom{n}{\dfrac{n-d}{2}}\binom{n-\dfrac{n-d}{2}}{\dfrac{n-d}{2}}=\binom{n}{\dfrac{n-d}{2}}\binom{\dfrac{n+d}{2}}{\dfrac{n-d}{2}}.$$

The first binomial coefficient is the number of ways to select (from $n$ games) the $\frac{1}{2}(n-d)$ games that the champ wins, and the second binomial coefficient is the number of ways to select (from the remaining games) the $\frac{1}{2}(n-d)$ games that the challenger wins. Then, for $n$ odd, the total probability of the tied match at the end of $n$ games is

$$C(0,n)=\sum_{d=1,\,3,\,\ldots,\,n}\binom{n}{\dfrac{n-d}{2}}\binom{\dfrac{n+d}{2}}{\dfrac{n-d}{2}}$$
$$[p(1-p-q)]^{(n-d)/2}\,q^d.$$

For the case of $n$ even, nothing changes in the above argument except that now $d$ must be an even number of drawn games. So, for $n$ even, the total probability of a tied match at the end of $n$ games is

$$C(0,n)=\sum_{d=0,\,2,\,\ldots,\,n}\binom{n}{\dfrac{n-d}{2}}\binom{\dfrac{n+d}{2}}{\dfrac{n-d}{2}}$$
$$[p(1-p-q)]^{(n-d/2)}\,q^d.$$

In Figure 10.1, I've shown the C-matrix of probabilities for a six-game match, and have indicated the directly calculable entries. Notice, in particular, that the entries below the diagonal are all zero; i.e., $C(k,n)=0$ for $k>n$. This is

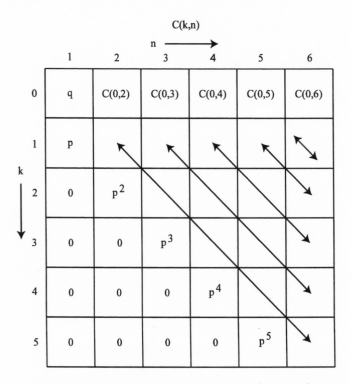

$C(k,n)$

$n \longrightarrow$

| | 1 | 2 | 3 | 4 | 5 | 6 |
|---|---|---|---|---|---|---|
| 0 | q | C(0,2) | C(0,3) | C(0,4) | C(0,5) | C(0,6) |
| 1 | p | | | | | |
| 2 | 0 | $p^2$ | | | | |
| 3 | 0 | 0 | $p^3$ | | | |
| 4 | 0 | 0 | 0 | $p^4$ | | |
| 5 | 0 | 0 | 0 | 0 | $p^5$ | |

$k$ (downward)

Figure 10.1 Probabilities for a Six-Game Chess Match

simply the physically obvious statement that the champ cannot be ahead by $k$ points unless at least $k$ games have been played. By comparing this partially filled-in matrix with the recurrence equation for $C$ you should be able to convince yourself that all the rest of the entries can now be calculated—if we do so in the proper order. That is, if we calculate the missing $C$ values along the diagonal paths shown in Figure 10.1, from left to right and starting with the lowest diagonal and moving upward, then each new $C$ value depends only on previously calculated values.

The MATLAB program **chess.m** (Program 20) performs all of these calculations for an $N$-game match, with a given

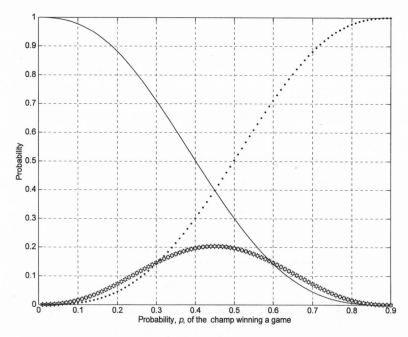

Figure 10.2 Win/Tie Probabilities for $N = 6$, $q = 0.1$ (tie is hexagrams, champ wins is dots, challenger wins is solid)

value for an individual game ending in a draw (the values of $N$ and $q$ are asked for by the program), and the following three plots show the probabilities asked for: i.e., plots (Figure 10.2) of the tied and won probabilities for the $N = 6$, $q = 0.1$ case. In addition, the plot (Figure 10.3) for the $N = 6$, $q = 0$ case shows that the probability of a tied match, for $p = \frac{1}{2}$, is just over 0.3, which agrees with the theoretical result of $20p^3(1-p)^3 = \frac{20}{2^6} = \frac{20}{64} = \frac{5}{6} = 0.3125$. Also, for the $N = 7$, $q = 0$ case, **chess.m** plots (Figure 10.4) the hexagrams of the "tied match" curve right on top of the horizontal axis, which agrees with the theoretical impossibility of a tied match for $N$ odd when $q = 0$.

Finally, the MATLAB program **ash.m** (Program 21) is **chess.m** slightly modified to run the special case of $p =$

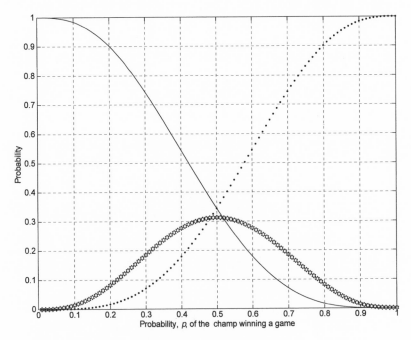

Figure 10.3  Win/Tie Probabilities for $N = 6$, $q = 0$ (tie is hexagrams, champ wins is dots, challenger wins is solid)

$q = \frac{1}{3}$, for any value of $N$. It runs very fast, too, since it is not stepping through many values of $p$ (all the plotting commands of **chess.m** have, of course, also been removed). When I ran **ash.m** for $N = 100$, it calculated the probability of a tied match to be 0.04877. Ash's asymptotic formula gives

$$\sqrt{\frac{3}{4\pi (100)}} = \frac{1}{20} \sqrt{\frac{3}{\pi}} = 0.04886.$$

This is reasonably good agreement. Although this does not prove that the coding for **chess.m** is error-free, I think it makes it very unlikely that there are errors.

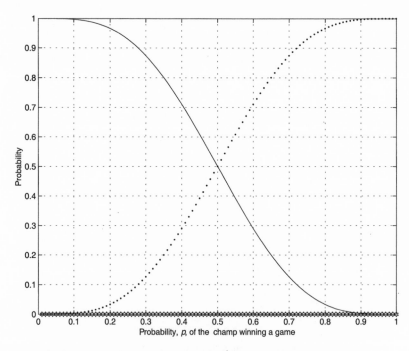

Figure 10.4 Win/Tie Probabilities for $N = 7$, $q = 0$ (Tie is hexagrams, champ wins is dots, challenger wins is solid)

## 11. A Different Slice of Probabilistic Pi

Suppose we've just tossed the needle and it has come to rest in some way on the tabletop. We are free to pick our coordinate system in any way we wish, of course, so let's pick it so that the needle is parallel with the x-axis and is in the upper half of the tabletop; i.e., the needle lies along the horizontal line $y = \text{constant} \geq 0$. So, from now on we are dealing only with a needle parallel to the x-axis, and we can do this with no loss in generality.

Now, let's be even more restrictive: Suppose that the needle does not stick out over the edge of the tabletop and, indeed, is right on the x-axis with its left end just at the edge. Then, imagine moving the needle in such a way that it both remains parallel to the x-axis and its left end continues to just touch the edge of the tabletop. This means, by definition, that the left end of the needle is moving along a circle with radius $r$ and centered on the middle of the tabletop.

Indeed, all of the points on the (rigid) needle must then be traveling on circular paths with radius $r$, but the centers of these paths are displaced along the x-axis. In particular, the midpoint of the needle will be orbiting the point $(a, 0)$; the equation of the path of the midpoint is $(x - a)^2 + y^2 = r^2$. This is shown in Figure 11.1 as Arc #1. A similar argument,

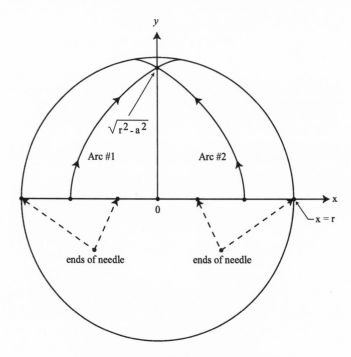

Figure 11.1 Needle-Drop Possibilities

for a needle on the x-axis with its right end just touching the tabletop edge, shows that its midpoint will travel along Arc #2, which has equation $(x+a)^2 + y^2 = r^2$.

Notice carefully that Arc #1 and Arc #2 cross on the y-axis, and it is there that both ends of the needle just touch the edge of the tabletop. The y value at this intersection point is $\sqrt{r^2 - a^2}$. We can now make what I think are three obvious observations:

1. If a needle lands with its midpoint to the left of Arc #1, then its left end sticks out over the edge of the table.

2. If a needle lands with its midpoint to the right of Arc #2, then its right end sticks out over the edge of the table.

3. If a needle lands with its midpoint between the two arcs, then

a. If the midpoint is below the intersection point of the two arcs, then neither end sticks out over the edge; and

b. If the midpoint is above the intersection point, then both ends stick out over the edge.

Figure 11.2 shows each of these three regions, and in particular takes advantage of their symmetry around the y-axis. Because of that symmetry, we can limit ourselves to the areas of the three regions that are in the first quadrant. So,

area of $I/\frac{1}{4}\pi r^2$ = probability that neither end of the needle sticks out over the edge of the tabletop;

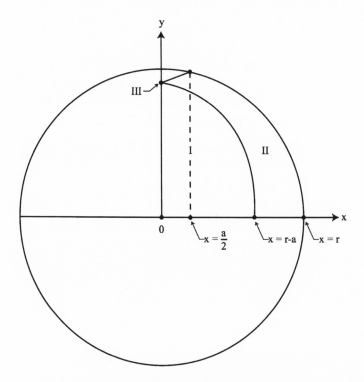

Figure 11.2  Needle-Drop Probabilities

area of II/$\frac{1}{4}$ $\pi$ $r^2$ = probability that one end (either right or left, but not both) sticks out over the edge of the tabletop;

area of III/$\frac{1}{4}$ $\pi$ $r^2$ = probability that both ends stick out over the edge of the tabletop.

All we have left to do, then, are standard freshman calculus problems of determining areas bounded by given curves. For example,

$$\text{area of I} = \int_0^{r-a} \sqrt{r^2 - (x+a)^2}\, dx = r \int_0^{r-a} \sqrt{1 - \left(\frac{x+a}{r}\right)^2}\, dx.$$

Make the change of variable $u = \frac{x+a}{r}$. Then $du = \frac{1}{r} dx$ and so

$$\text{area of I} = r \int_{\frac{a}{r}}^{1} \sqrt{1-u^2}\, r\, du = r^2 \int_{\frac{a}{r}}^{1} \sqrt{1-u^2}\, du$$

$$= r^2 \left[ \frac{u}{2} \sqrt{1-u^2} + \frac{1}{2} \sin^{-1}(u) \ \Big|_{\frac{a}{r}}^{1} \right.$$

$$= r^2 \left[ \frac{1}{2} \sin^{-1}(1) - \frac{a}{2r} \sqrt{1 - \left(\frac{a}{r}\right)^2} \right.$$

$$\left. - \frac{1}{2} \sin^{-1}\left(\frac{a}{r}\right) \right].$$

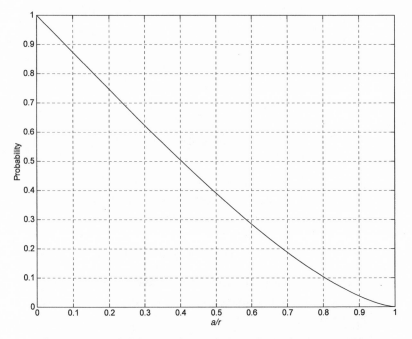

Figure 11.3    Probability Neither End of Needle Sticks Out over Edge

Thus,

$$P \text{ (neither end sticks out over edge)} =$$

$$\frac{r^2 \left[ \dfrac{\pi}{4} - \dfrac{a}{2r} \sqrt{1 - \left(\dfrac{a}{r}\right)^2} - \dfrac{1}{2} \sin^{-1}\left(\dfrac{a}{r}\right) \right]}{\dfrac{1}{4} \pi r^2}$$

$$= 1 - \frac{2}{\pi} \left[ \left(\frac{a}{r}\right) \sqrt{1 - \left(\frac{a}{r}\right)^2} + \sin^{-1}\left(\frac{a}{r}\right) \right].$$

If we do the same for the area of III, we have

$$\text{area of III} = \int_{0}^{\frac{1}{2} a} \left[ \sqrt{r^2 - x^2} - \sqrt{r^2 - (x - a)^2} \right] dx$$

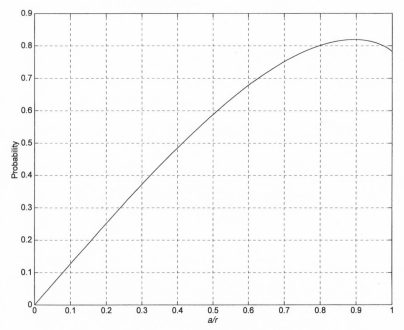

Figure 11.4   Probability One End of Needle Sticks Out over Edge

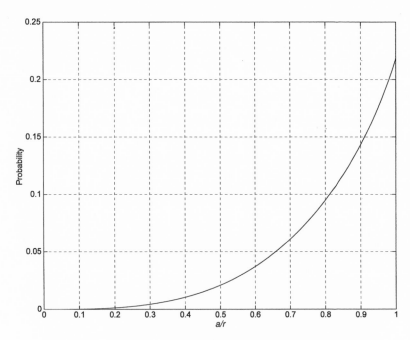

Figure 11.5 Probability Both Ends of Needle Stick Out over Edge

which, after a bit of (not very difficult) manipulation, results in

$$P \text{ (both ends stick out over edge)}$$

$$= \frac{2}{\pi} \left[ \left( \frac{a}{r} \right) \sqrt{1 - \frac{1}{4} \left( \frac{a}{r} \right)^2} + 2 \sin^{-1} \left\{ \frac{1}{2} \left( \frac{a}{r} \right) \right\} \right.$$

$$\left. - \left( \frac{a}{r} \right) \sqrt{1 - \left( \frac{a}{r} \right)^2} - \sin^{-1} \left( \frac{a}{r} \right) \right].$$

We could also do the integral for II (and that's a good exercise for you), but I'll simply take the probability that one end sticks out over the edge as 1 minus the sum of the above two probabilities. Notice that the probabilities are functions only of the ratio of $a$ and $r$; Figures 11.3, 11.4, and 11.5, produced by **needle.m** (Program 22), show how the three probabilities vary over the interval $0 \le a/r \le 1$.

*Historical note:* The idea of dropping an object at random to estimate $\pi$ fascinates most people; to see how one man built a machine to do this, see two papers by A. L. Clark, "An Experimental Investigation of Probability," *The Canadian Journal of Research* (vol. 9, October 1933, pp. 402–414, and vol. 11, November 1934, pp. 658–664).

## 12. When Negativity Is a No-No

Since $Z = \frac{X}{X-Y}$ with $X$ and $Y$ both varying (independently) over 0 to 1, then it is clear that $-\infty < Z < \infty$. To find the probability density function of $Z$, I will use the standard method of first computing the distribution function $F_Z(z)$ $= P(Z \le z)$ and then differentiating, i.e.,

$$f_Z(z) = \frac{d}{dz} F_Z(z).$$

This is not a technically difficult problem, but it does demand being very careful with inequalities. So, we have

$$F_Z(z) = P(Z \le z) = P\left( \frac{X}{X-Y} \le z \right)$$
$$= \begin{bmatrix} P(X \le zX - zY) \text{ if } X - Y > 0 \\ P(X \ge zX - zY) \text{ if } X - Y < 0 \end{bmatrix}.$$

Or,

$$F_Z(z) = \begin{bmatrix} P\{zY \le (z-1)X\} \text{ if } X - Y > 0 \\ P\{zY \ge (z-1)X\} \text{ if } X - Y < 0 \end{bmatrix}.$$

Notice, carefully, that when multiplying through an inequality by a negative quantity, we must be sure to *reverse* the sense of the inequality. (That is what my students consistently failed do.)

We can write the last statement, then, as follows, where once again we have to be careful to reverse the sense of an inequality if we divide through the inequality by a negative quantity:

$$F_Z(z) = \begin{bmatrix} P(Y \leq \frac{z-1}{z} X) \text{ if } X - Y > 0 \text{ and } z > 0 \\ P(Y \geq \frac{z-1}{z} X) \text{ if } X - Y > 0 \text{ and } z < 0 \\ P(Y \geq \frac{z-1}{z} X) \text{ if } X - Y < 0 \text{ and } z > 0 \\ P(Y \leq \frac{z-1}{z} X) \text{ if } X - Y < 0 \text{ and } z < 0 \end{bmatrix}.$$

Alternatively, we can write the above four statements as

(A)  If $z < 0$ then $F_Z(z) =$

$$\begin{bmatrix} P(Y \leq \frac{z-1}{z} X) \text{ if } X - Y < 0 \\ P(Y \geq \frac{z-1}{z} X) \text{ if } X - Y > 0 \end{bmatrix}$$

(B)  If $z > 0$ then $F_Z(z) =$

$$\begin{bmatrix} P(Y \geq \frac{z-1}{z} X) \text{ if } X - Y < 0 \\ P(Y \leq \frac{z-1}{z} X) \text{ if } X - Y > 0 \end{bmatrix}$$

Using conditional notation—i.e., $P(U \mid V) =$ probability of event $U$ given event $V$—then (A) and (B) become

(A′) If $z < 0$ then

$$\left[ \begin{array}{l} F_Z(z|X - Y < 0) = P(Y \leq \dfrac{z-1}{z} X | X - Y < 0) \\[3mm] F_Z(z|X - Y > 0) = P(Y \geq \dfrac{z-1}{z} X | X - Y > 0) \end{array} \right]$$

(B′) If $z > 0$ then

$$\left[ \begin{array}{l} F_Z(z|X - Y < 0) = P(Y \geq \dfrac{z-1}{z} X | X - Y < 0) \\[3mm] F_Z(z|X - Y > 0) = P(Y \leq \dfrac{z-1}{z} X | X - Y > 0) \end{array} \right].$$

Since we can write

$$F_Z(z) = F_Z(z|X - Y < 0)P(X - Y < 0) \\ + F_Z(z|X - Y > 0)P(X - Y > 0),$$

and since $P(X - Y < 0) = P(X - Y > 0) = \frac{1}{2}$, as shown in Figure 12.1, then we have for (A′), i.e., for $z < 0$:

$$F_Z(z) = \frac{1}{2} \left[ P(Y \leq \frac{z-1}{z} X | X - Y < 0) \right. \\ \left. + P(Y \geq \frac{z-1}{z} X | X - Y > 0) \right].$$

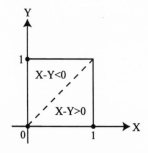

Figure 12.1 Equal Probabilities for $X - Y < 0, X - Y > 0$

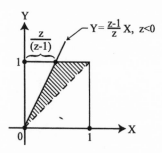

Figure 12.2 Probability for $Y \leq (z - 1/z) X$ for $z < 0$

Now, since $\frac{z-1}{z} = 1 - \frac{1}{z} > 1$ for $z < 0$, then the line $Y = \frac{z-1}{z} X$ lies above the diagonal line $Y = X$ that divides the $(X, Y)$ sample space into two equal (probability) parts, and Figure 12.2 shows the shaded region in sample space that satisfies the first term in the above expression for $F_Z(z)$. That is, the shaded region is where $Y \leq \frac{z-1}{z} X$ for $z < 0$, given that $X - Y < 0$. By inspection, you can also see that the second term is zero, i.e., for $z < 0$,

$$P(Y \geq \tfrac{z-1}{z} X | X - Y > 0) = 0$$

because the region $Y \geq \frac{z-1}{z} X$ and the $X - Y > 0$ half of the sample space do not overlap.

So, by geometry, we have the area of the shaded region in Figure 12.2 (which, when divided by the area of the $X - Y < 0$ half of the sample space, $\frac{1}{2}$, gives us the probability of the shaded region since $X$ and $Y$ are uniform) and thus

$$P(Y \leq \frac{z-1}{z} X | X - Y < 0) = \frac{\dfrac{1}{2} - \dfrac{1}{2} \times \dfrac{z}{z-1}}{\dfrac{1}{2}}$$

$$= 1 - \frac{z}{z-1}, \; z < 0.$$

Therefore, for $z < 0$,

$$F_Z(z) = \frac{1}{2}\left[1 - \frac{z}{z-1}\right] = \frac{1/2}{1-z}, \, z < 0.$$

Our problem is now half solved. We must now find $F_Z(z)$ for $z > 0$, for which I will turn to (B').

As before, we can write, for $z > 0$,

$$F_Z(z) = F_Z\,(z|X - Y < 0)P(X - Y < 0)$$

$$+ F_Z\,(z|X - Y > 0)P(X - Y > 0)$$

$$= \frac{1}{2}\left[P(Y \le \frac{z-1}{z}\,X|X - Y < 0)\right.$$

$$\left. + P(Y \le \frac{z-1}{z}\,X|X - Y > 0)\right].$$

What we do next is just a bit different from before, because there are two cases we must consider: $\frac{z-1}{z} = 1 - \frac{1}{z}$ and so

(a) $1 - \frac{1}{z} < 0$ for $0 < z < 1$

(b) $0 < 1 - \frac{1}{z} < 1$ for $z > 1$.

Case (a) is shown in Figure 12.3, where the line $Y = \frac{z-1}{z}$ $X$ is drawn with a negative slope. The picture immediately tells us that $P(Y \ge \frac{z-1}{z} X|X - Y < 0) = 1$, since all values of $Y$ in the sample space region defined by $X - Y < 0$ lie above the line, and also that $P(Y \le \frac{z-1}{z} X|X - Y > 0) = 0$ since no value of $Y$ in the sample space region defined by $X - Y > 0$ lies below the line. So, we have the interesting result of

$$F_Z(z) = \frac{1}{2}\,[1 + 0] = \frac{1}{2}, \, 0 < z < 1.$$

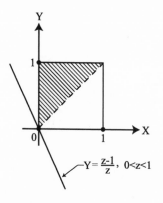

Figure 12.3  Case (a) (see text)

Case (b) is shown in Figure 12.4, from which it is geometrically clear that

$$P\left(Y \geq \frac{z-1}{z} X \mid X - Y < 0\right) = 1$$

and that

$$P\left(Y \leq \frac{z-1}{z} X \mid X - Y > 0\right) = \frac{\dfrac{1}{2}\dfrac{z-1}{z}}{\dfrac{1}{2}} = \frac{z-1}{z}.$$

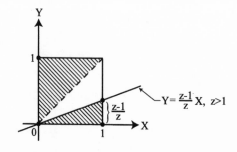

Figure 12.4  Case (b) (see text)

Thus,

$$F_Z(z) = \frac{1}{2}\left[1 + \frac{z-1}{z}\right] = \frac{2z-1}{2z} = 1 - \frac{1}{2z}, \; z > 1.$$

So, in summary, we have

$$F_Z(z) = P(Z \leq z) = \begin{bmatrix} \dfrac{\frac{1}{2}}{1-z}, \; z < 0 \\[2mm] \dfrac{1}{2}, \; 0 < z < 1 \\[2mm] 1 - \dfrac{1}{2z}, \; z > 1 \end{bmatrix}$$

Observe that $F_Z(-\infty) = 0$, $F_Z(\infty) = 1$, and $F_Z(z)$ increases monotonically with increasing $z$, three properties that all probability distribution functions must possess.

We can now find the probability density function $f_Z(z)$ by differentiating $F_Z(z)$, which gives

$$f_Z(z) = \begin{bmatrix} \dfrac{1}{2(1-z)^2}, \; z < 0 \\[2mm] 0, \; 0 < z < 1 \\[2mm] \dfrac{1}{2z^2}, \; z > 1 \end{bmatrix}.$$

Figure 12.5 shows what a simulation of $Z$ says that $f_Z(z)$ looks like. (See the MATLAB program **z.m.** [Program 23]) Notice that $f_Z(z) > 0$ for $-\infty < z < \infty$; i.e., the density function is never negative (which is as it should be.) It is interesting to note that, since $f_Z(z) = 0$ for $0 < z < 1$, there are no values of $Z$ between 0 and 1. This is actually obvious from the definition of $Z$, once you think about it, but I didn't see it until I formally worked all the way through the problem. That is, suppose there could be a $Z$ in the interval 0 to 1; call its value $z$. Then, solving for Y, we have

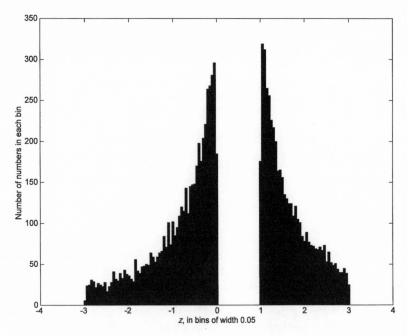

Fig\gamma re 12.5  Histogram of $Z = X/(X-Y)$

$$Y = -\frac{1-z}{z} X.$$

But, since $0 < z < 1$, then $-\frac{1-z}{z} < 0$, so Y and X would have opposite signs. But that is impossible, since both X and Y are always between 0 and 1.

## 13. The Power of Randomness

To start, I'll write the probability distribution function of $Z = X^Y$ as

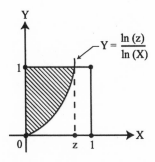

Figure 13.1 Shaded Region Shows Where Inequality Is Satisfied

$$F_Z(z) = P(Z \leq z) = P(X^Y \leq z) = P[Y \ln(X) \leq \ln(z)].$$

Dividing through the last inequality by $\ln(X)$ and remembering that it is always non-positive because $0 \leq X \leq 1$, we must reverse the sense of the inequality. So,

$$F_Z(z) = P\left[Y \geq \frac{\ln(z)}{\ln(X)}\right].$$

It is clear that, by its definition, $0 \leq Z \leq 1$, so $\ln(z) \leq 0$ is the case of interest for us. This means that $\ln(z)/\ln(X)$ is always non-negative; i.e., $Y \geq 0$, which is consistent with $0 \leq Y \leq 1$. I've sketched the new inequality, plotted on the sample space for $X$ and $Y$ (the unit square), in Figure 13.1. The shaded region represents where the inequality is satisfied in the sample space.

Now, from the geometric interpretation of the probability of a region of sample space (if the random variables of interest are uniformly distributed) as the ratio of the area of that region to the area of the entire sample space, we immediately have

$$F_Z(z) = z - \int_0^z \frac{\ln(z)}{\ln(x)}\, dx = z - \ln(z) \int_0^z \frac{dx}{\ln(x)}.$$

Notice that this expression does satisfy the endpoint requirements of any distribution function for a random variable that takes values from the unit interval; i.e., $F_Z(0) = 0$ and $F_Z(1) = 1$.

To find the probability density function for $Z$ we need only differentiate the distribution, i.e.,

$$f_Z(z) = \frac{d}{dz} F_Z(z).$$

This requires us to differentiate an integral, a calculation you can do if you remember Leibniz's rule from calculus, i.e.,

$$\frac{d}{dz} \int_{\alpha(z)}^{\beta(z)} g(z, u)\, du = \int_{\alpha(z)}^{\beta(z)} \frac{dg}{dz}\, du + g(z, \beta(z)) \frac{d\beta}{dz} - $$
$$g(z, \alpha(z)) \frac{d\alpha}{dz}.$$

So,

$$f_Z(z) = 1 - \left[ \ln(z) \frac{1}{\ln(z)} + \frac{1}{z} \int_0^z \frac{dx}{\ln(x)} \right] = -\frac{1}{z} \int_0^z \frac{dx}{\ln(x)},$$
$$0 \le z \le 1.$$

To express this more neatly, change the dummy integration variable to $u = -\ln(x)$, i.e., to $x = e^{-u}$. Then, $\frac{du}{dx} = -\frac{1}{x}$, or $dx = -x\, du = -e^{-u} du$. Our pdf expression then becomes

$$f_Z(z) = -\frac{1}{z} \int_\infty^{-\ln(z)} \frac{-e^{-u}}{-u}\, du = \frac{1}{z} \int_{-\ln(z)}^\infty \frac{e^{-u}}{u}\, du,$$
$$0 \le z \le 1.$$

Don't forget: Since $0 \leq z \leq 1$, then the lower limit on the integral is never negative.

Now, the exponential integral function is defined as

$$Ei(x) = \int\limits_x^\infty \frac{e^{-u}}{u} \, du.$$

So, we immediately have

$$f_Z(z) = \frac{1}{z} \, Ei \, [- \ln(z)], \, 0 \leq z \leq 1.$$

The MATLAB program **xpowery.m** (Program 24) plots this theoretical expression in Figure 13.2, while **xyhisto.m** (Program 25) constructs a histogram of 20,000 $X^Y$ values in Figure 13.3. The two plots do indeed "look the same."

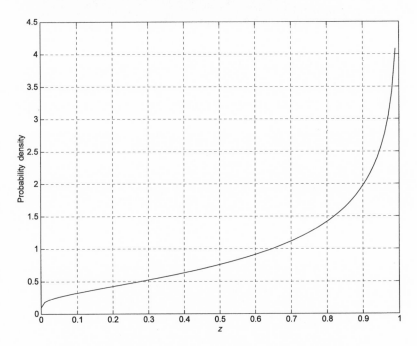

Figure 13.2  PDF of $Z = X^Y$, with $X$ and $Y$ Uniform from 0 to 1

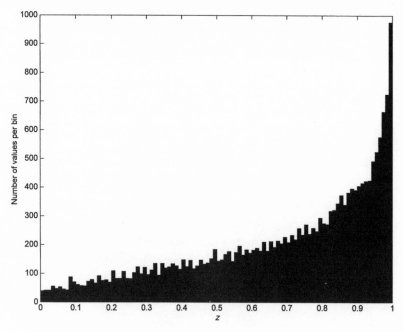

Figure 13.3 Histogram of $X^Y$ (20,000 values)

## 14. The Random Radio

The solutions to the totally random quadratic $Ax^2 + Bx + C = 0$ are given by

$$x = \frac{-B \pm \sqrt{B^2 - 4AC}}{2},$$

and so are real if and only if $B^2 \geq 4AC$. Define two new random variables, $X = B^2$ and $Y = 4AC$. $X$ and $Y$ are clearly independent because they are functions of entirely different, independent variables. $X$ varies from 0 to 1, while $Y$ varies from 0 to 4. The answer to our problem, then, is just $P(X \geq Y)$, the probability that $X \geq Y$. The sample space for

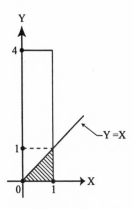

Figure 14.2  Pairs of X and Y That Result in Real Roots

X and Y, with the inequality $X \geq Y$ plotted on it, looks like Figure 14.2.

The shaded region is the collection of all pairs of values for X and Y that result in real roots, and so the answer to our problem is the probability of the shaded region. But now, however, we cannot simply divide the shaded area by the area of sample space, because this is no longer a problem in geometric probability (because neither X nor Y is uniform). Instead, we must integrate the joint probability density function for X and Y, $f_{X,Y}(x, y)$, over the shaded region. Since X and Y are at least independent, then we do have the joint density function as the product of the marginal density functions for X and Y, $f_X(x)$ and $f_Y(y)$, respectively. That is, $f_{X,Y}(x, y) = f_X(x)f_Y(y)$. So, we must next find $f_X(x)$ and $f_Y(y)$. The density function for X is straightforward.

The probability distribution function for X, $F_X(x)$, is

$$F_X(x) = P(X \leq x) = P(B^2 \leq x) = P(-\sqrt{x} \leq B \leq \sqrt{x})$$
$$= P(0 \leq B \leq \sqrt{x}) = \sqrt{x}$$

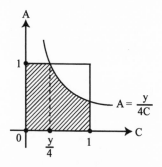

$A = \dfrac{y}{4C}$

Figure 14.3   Sample Space for A and C

because X is uniform from 0 to 1 (and so X never takes on negative values). Thus,

$$f_X(x) = \frac{d}{dx} F_X(x) = \frac{1}{2\sqrt{x}},\ 0 \le x \le 1.$$

We have to do just a bit more work for $f_Y(y)$.

As with X, we first find the distribution function for Y:

$$F_Y(y) = P(Y \le y) = P(4AC \le y) = P\!\left(A \le \frac{y}{4C}\right).$$

The sample space for A and C, with the inequality $A \le \frac{y}{4C}$ plotted on it, is shown in Figure 14.3. $F_Y(y)$ is the probability of the shaded region, which, since A and C are uniform, is easily found with the geometric probability concept of the ratio of areas. So, since the entire sample space has unity area,

$$F_Y(y) = \frac{y}{4} + \int_{\frac{y}{4}}^{1} \frac{y}{4C}\, dC = \frac{y}{4} + \frac{y}{4}\left(\ln C\,\Big|_{\frac{y}{4}}^{1}\right).$$

$$= \frac{y}{4} - \frac{y}{4}\ln\!\left(\frac{y}{4}\right).$$

Thus,

$$f_Y(y) = \frac{d}{dy} F_Y(y) = -\frac{1}{4} \ln\left(\frac{y}{4}\right), \ 0 \le y \le 4.$$

Thus, the joint density function for X and Y is

$$f_{X,Y}(x, y) = -\frac{\ln\left(\dfrac{y}{4}\right)}{8\sqrt{x}}, \ \begin{matrix} 0 \le x \le 1 \\ 0 \le y \le 4. \end{matrix}$$

Now, returning to the shaded region in the X, Y sample space of Figure 14.2, we have for our answer

$$P(X \ge Y) = \int_0^1 \left( \int_0^x -\frac{\ln\left(\dfrac{y}{4}\right)}{8\sqrt{x}} \, dy \right) dx$$

$$= \int_0^1 -\frac{1}{8\sqrt{x}} \left\{ \int_0^x \ln\left(\frac{y}{4}\right) dy \right\} dx.$$

In the inner integral, change the variable to $u = \frac{1}{4}y$, so $dy = 4du$. Then,

$$P(X \ge Y) = -\frac{1}{8} \int_0^1 \frac{1}{\sqrt{x}} \left[ \int_0^{\frac{1}{4}x} \ln(u) 4du \right] dx$$

$$= -\frac{1}{2} \int_0^1 \frac{1}{\sqrt{x}} \left[ u\ln(u) - u \Big|_0^{\frac{1}{4}x} \right] dx.$$

$$= -\frac{1}{2} \int_0^1 \frac{1}{\sqrt{x}} \left[ \frac{1}{4}x \ln\left(\frac{1}{4}x\right) - \frac{1}{4}x \right] dx$$

$$= -\frac{1}{8} \int_0^1 \sqrt{x} \ln\left(\frac{1}{4}x\right) dx + \frac{1}{8} \int_0^1 \sqrt{x} \, dx.$$

The second integral is easy to do, and it should be no trouble for you to verify that

$$\frac{1}{8} \int_0^1 \sqrt{x} \, dx = \frac{1}{12}.$$

Thus,

$$P(X \geq Y) = \frac{1}{12} - \frac{1}{8} \int_0^1 \sqrt{x} \ln\left(\frac{1}{4}x\right) dx.$$

The last integral, while still elementary, does require just a bit more effort. Change the dummy integration variable to $u = \frac{1}{2}x^{1/2}$. Then, $du = \frac{1}{4}x^{-1/2} \, dx$, or $dx = 4x^{1/2} \, du = 4 \cdot 2u \, du = 8u \, du$. That is, $\sqrt{x} = 2u$, so $u^2 = \frac{1}{4}x$. That gives us

$$P(X \geq Y) = \frac{1}{12} - \frac{1}{8} \int_0^{\frac{1}{2}} 2u \ln(u^2)8u \, du$$

$$= \frac{1}{12} - 2 \int_0^{\frac{1}{2}} u^2 \ln(u^2) \, du$$

$$= \frac{1}{12} - 4 \int_0^{\frac{1}{2}} u^2 \ln(u) \, du$$

$$= \frac{1}{12} - 4 \left\{ \frac{1}{3}u^3 \ln(u) - \frac{1}{9} u^3 \Big|_0^{\frac{1}{2}} \right.$$

$$= \frac{1}{12} - 4 \left\{ \frac{1}{3} \cdot \frac{1}{8} \ln\left(\frac{1}{2}\right) - \frac{1}{9} \cdot \frac{1}{8} \right\}$$

$$= \frac{1}{12} - \frac{1}{6} \ln\left(\frac{1}{2}\right) + \frac{1}{18} = \frac{1}{12} + \frac{1}{6} \ln(2) + \frac{1}{18}$$

$$= \frac{5 + 6 \ln(2)}{36} = 0.25441341898221 \ldots,$$

the probability that the totally random quadratic equation has real solutions, which is more than three times greater than the probability that the partially random quadratic equation has real solutions.

That is the solution when we make no independence assumptions beyond the originally given independence of

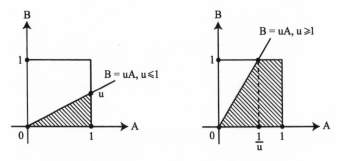

Figure 14.4 Probabilities for $0 \le u \le 1$ and $1 \le u \le \infty$

A, B, and C. Suppose, now, that we divide through the totally random quadratic by A to get $x^2 + Ux + V = 0$, where $U = B/A$, and $V = C/A$ (which will have real solutions if and only if $U^2 \ge 4V$). U and V have the same probability density function, of course, since B and C are both uniform from 0 to 1. (Both U and V vary from 0 to ∞.) Thus, it will be sufficient to find the density for just one of them; let's do U. For the distribution of U, we have

$$F_U(u) = P(U \le u) = P\left(\frac{B}{A} \le u\right) = P(B \le uA).$$

Figure 14.4 shows the geometry of this probability for the two cases of $0 \le u \le 1$ and $1 \le u < \infty$. Since A and B are uniform and independent, we have (by geometric probability, by inspection),

$$F_U(u) = \frac{1}{2}u, \ 0 \le u \le 1$$

$$= 1 - \frac{1}{2u}, \ u \ge 1.$$

Thus,

$$f_U(u) = \frac{d}{du} F_U(u) = \frac{1}{2}, \ 0 \le u \le 1$$

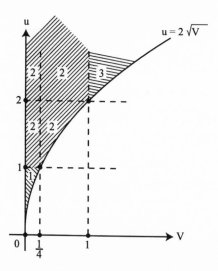

Figure 14.5 Points in Sample Space with Real Solutions

$$= \frac{1}{2u^2}, \ u \geq 1.$$

And, as I pointed out earlier, $V$ has the identical statistical behavior; therefore,

$$f_V(v) = \frac{1}{2}, \ 0 \leq v \leq 1$$

$$= \frac{1}{2v^2}, \ v \geq 1.$$

Now we make our major assumption: that $U$ and $V$ are independent. (This will turn out to be wrong, but we don't actually know that yet.) The sample space for $U$ and $V$ is the entire first quadrant of Figure 14.5, where the shaded region (infinite in extent) represents the set of all points in the sample space associated with real solutions. To find the probability of that region, $P(U \geq 2\sqrt{V})$, we integrate the joint probability density function of $U$ and $V$ over it. And by our assumption, we take that joint density to be

$$f_{U,V}(u, v) = f_U(u)\, f_V(v),$$

where we will use the appropriate form for each marginal density function, depending on where we happen to be in the first quadrant while doing our integral.

In the notation of Figure 14.5, in fact, we have:

$$f_{U,V}(u, v) = \frac{1}{2} \cdot \frac{1}{2} = \frac{1}{4} \text{ in Region 1,}$$

$$f_{U,V}(u, v) = \frac{1}{2} \cdot \frac{1}{2u^2} = \frac{1}{4u^2} \text{ in Region 2,}$$

and

$$f_{U,V}(u, v) = \frac{1}{2u^2} \cdot \frac{1}{2v^2} = \frac{1}{4u^2 v^2} \text{ in Region 3.}$$

Thus, the final calculation we must do is

$$P(U \geq 2\sqrt{V}) = \int_0^{\frac{1}{4}} \left( \int_{2\sqrt{v}}^1 \frac{1}{4}\, du \right) dv$$

$$+ \int_0^{\frac{1}{4}} \left( \int_1^\infty \frac{1}{4u^2}\, du \right) dv + \int_{\frac{1}{4}}^1 \left( \int_{2\sqrt{v}}^\infty \frac{1}{4u^2}\, du \right) dv$$

$$+ \int_1^\infty \left( \int_{2\sqrt{v}}^\infty \frac{1}{4u^2 v^2}\, du \right) dv.$$

All four integrals are very easy to do (I'll let you fill in the details), with values of $\frac{1}{48}$, $\frac{1}{16}$, $\frac{1}{8}$, and $\frac{1}{12}$, respectively. So,

$$P(U \geq 2\sqrt{V}) = \frac{1}{48} + \frac{1}{16} + \frac{1}{8} + \frac{1}{12} = \frac{14}{48} = \frac{7}{24}$$

$$= 0.291666\dots,$$

the probability the totally random quadratic has real solutions if $U$ and $V$ are independent; i.e., if $B/A$ and $C/A$ are independent. This value does not agree with the first calculation, however, so $B/A$ and $C/A$ are not independent and this numerical result is wrong. In fact, when I ran a MATLAB program that checked 100,000 triples of random numbers in the inequality $B^2 \geq 4AC$, the result was 0.25424, which compares well with the first calculation.

A final comment: If you are familiar with the elegant Jacobian determinant method for calculating the joint probability density function of multiple random variables that are functions of yet other random variables, then you might try your hand at showing that the joint pdf of $Z = B/A$ and $W = C/A$ is

$$
f_{Z,W}(z, w) = \begin{cases} \dfrac{1}{3} & \text{if } z < 1,\ w < 1 \\[2mm] \dfrac{1}{3z^3} & \text{if } z > 1 \text{ and } z > w \\[2mm] \dfrac{1}{3w^3} & \text{if } w > 1 \text{ and } z < w. \end{cases}
$$

The condition for real roots is $Z^2 - 4W \geq 0$; if you integrate this joint pdf over that part of the first quadrant in the $Z, W$ plane, you will find, once again, that the probability of real roots to the totally random quadratic is $\dfrac{5 + 6 \ln(2)}{36}$. This provides a nice confirmation of the analysis presented earlier.

## 15. An Inconceivable Difficulty

If $p = 0$ (only girls) or $p = 1$ (only boys), then the average number of children in a family is clear: 2, *always*. Most students intuitively feel that as $p$ varies smoothly from 0 to 1, the average number of children should also vary smoothly from 2 up to some maximum (at $p = \frac{1}{2}$, by symmetry) and then back down to 2. This feeling is, in fact, incorrect. What actually occurs is that when $p \neq 0$ and $p \neq 1$, then the average number of children is 3, *always*. When $p$ reaches either 0 or 1, then the average number of children discontinuously drops to 2.

To see that this rather curious behavior of the average value is what occurs, let's write out the two sample points of length $n$ ($n \geq 2$) that satisfy the given conditions of the family planning rule:

$$B\,G\,G\ldots G\,B$$
$$G\,B\,B\ldots B\,G.$$

The first sample point has probability $p^2 (1-p)^{n-2}$ and the second one has probability $(1-p)^2 p^{n-2}$. So, the average number of children is

$$C = \sum_{n=2}^{\infty} \left[ p^2 (1-p)^{n-2} + (1-p)^2 p^{n-2} \right] n$$

$$= \frac{p^2}{(1-p)^2} \sum_{n=2}^{\infty} (1-p)^n\, n + \frac{(1-p)^2}{p^2} \sum_{n=2}^{\infty} p^n n.$$

That is, $C$ has the form of

$$C = \frac{p^2}{(1-p)^2} S_1 + \frac{(1-p)^2}{p^2} S_2$$

where the two sums $S_1$ and $S_2$ have the general form of

$$S = \sum_{n=2}^{\infty} a^n n\,.$$

So, $$S = 2a^2 + 3a^3 + 4a^4 + \ldots;$$

therefore, $$aS = 2a^3 + 3a^4 + \ldots;$$

which means that $\quad S - aS = S(1 - a) = 2a^2 + a^3 + a^4$

$$+ \ldots = a^2 + T$$

where $\quad\quad\quad\quad\quad T = a^2 + a^3 + a^4 + \ldots.$

But then $\quad\quad\quad\quad aT = a^3 + a^4 + \ldots,$

so $\quad\quad\quad\quad\quad T - aT = T(1 - a) = a^2$

or, $\quad\quad\quad\quad\quad T = \dfrac{a^2}{1 - a}.$

Thus, $\quad\quad\quad S(1 - a) = a^2 + \dfrac{a^2}{1 - a} = \dfrac{2a^2 - a^3}{1 - a}$

or, at last, $\quad\quad S = \dfrac{a^2 (2 - a)}{(1 - a)^2}.$

For the sum $S_1$ we have $a = 1 - p$, and for the sum $S_2$, we have $a = p$. Thus,

$$C = \frac{p^2}{(1-p)^2} \cdot \frac{(1-p)^2 (1+p)}{p^2} + \frac{(1-p)^2}{p^2} \cdot \frac{p^2 (2-p)}{(1-p)^2}$$

$$= 1 + p + 2 - p = 3.$$

This result fails, of course, for the cases of $p = 0$ and $p = 1$ (where, we already know, the answer is $C = 2$) because for those two special values of $p$, we have division by zero occurring in $S_2$ or $S_1$, respectively.

The MATLAB program **kids.m** (Program 26) simulates this family planning rule for 10,000 families. Running it three times, for various values of $p$, produced the following results, which confirm the theory.

|  | | $c$ | |
|---|---|---|---|
| 0 | 2 | 2 | 2 |
| .01 | 3.0654 | 2.9629 | 3.2228 |
| $p$  .3 | 3.0454 | 2.996 | 3.0117 |
| .7 | 3.0013 | 2.9818 | 2.9825 |
| 1 | 2 | 2 | 2 |

The puzzling aspect of this problem are the discontinuities in C at $p = 0$ and at $p = 1$. Why, students ask, does changing $p$ from 0 to .01 make C jump from 2 to 3? In fact, with $p = .01$, most families still do have just two children. But now with $p = .01$, there is a small chance of having substantially larger numbers of children. That raises the *average* to 3, even though most families have just two children. This is an example of a problem for which the average value is not a very informative number.

## 16. The Unsinkable Tub Is Sinking! How to Find Her, Fast

Assign $n$ search boats to Island #1 (and thus $N - n$ search boats to Island #2). If $P$ denotes the total probability of finding the *Unsinkable Tub*, then we have

$$P = \text{Prob } [\textit{Unsinkable Tub} \text{ is at Island \#1}] \cdot \text{Prob} \begin{Bmatrix} \text{at least one} \\ \text{search boat} \\ \text{finds the} \\ \textit{Tub} \end{Bmatrix}$$

$$+$$

Prob [*Unsinkable Tub* is at Island #2] · Prob $\left\{ \begin{array}{l} \text{at least one} \\ \text{search boat} \\ \text{finds the } Tub \end{array} \right\}$.

Thus,

$$P = p_1 \left[ 1 - (1 - p_s)^n \right] + p_2 \left[ 1 - (1 - p_s)^{N-n} \right]$$

$$= p_1 \left[ 1 - e^{n\ln(1 - p_s)} \right] + p_2 \left[ 1 - e^{(N-n)\ln(1 - p_s)} \right]$$

$$= p_1 \left[ 1 - e^{n\ln(1 - p_s)} \right] + p_2 \left[ 1 - e^{N\ln(1 - p_s)} \, e^{-n\ln(1 - p_s)} \right].$$

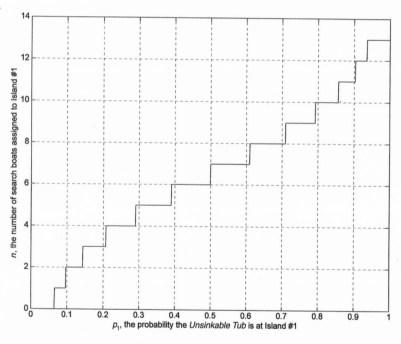

Figure 16.1 Probability of Finding the *Unsinkable Tub*, $N = 13$, $p_s = 0.2$

To find the value of $n$ that maximizes $P$, let's assume that $n$ is a continuous variable (although it's really a discrete variable, of course, equal to integer values), and then differentiate. So,

$$\frac{dP}{dn} = - p_1 \ln(1 - p_s) \, e^{n\ln(1-p_s)}$$

$$+ p_2 \, e^{N\ln(1-p_s)} \, e^{-n\ln(1-p_s)} \ln(1 - p_s).$$

Setting this equal to zero to locate the maximum of $P$, it quickly reduces to

$$(1 - p_s)^{2n} = \frac{p_2}{p_1} \, (1 - p_s)^N,$$

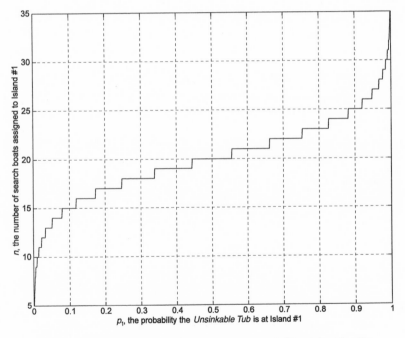

Figure 16.2 Probability of Finding the *Unsinkable Tub*, $N = 40$, $p_s = 0.2$

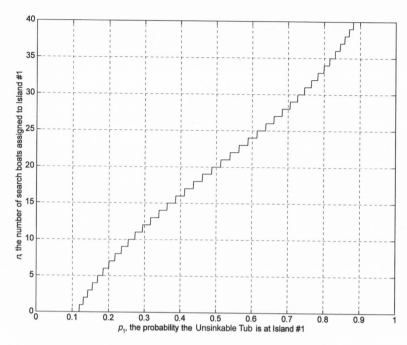

Figure 16.3 Probability of Finding the *Unsinkable Tub*, $N = 40$, $p_s = 0.5$

or, solving for $n$,

$$n = \frac{1}{2} \left[ N + \frac{\ln\left(\dfrac{p_2}{p_1}\right)}{\ln(1 - p_s)} \right].$$

This solution for $n$ will, in general, not be an integer. So, what we need to do is to evaluate $P$ for the two integers on either side of our solution value and select the one that gives the larger $P$. The two special cases we need to be alert for are $n < 0$ and $n > N$; in those two cases, we simply set $n = 0$ and $n = N$, respectively. The program **tub.m** (Program 27) performs all the calculations, for any given values of $N$ and $p_s$, for $p_1$ in the interval $0.001 \leq p_1 \leq 0.999$. The results asked for in the problem statement are given in Figures 16.1, 16.2, and 16.3

## 17. A Walk in the Garden

The MATLAB program **paths.m** (Program 28) performs the simulation sketched in the problem statement. Figure 17.2 shows the results, which include a couple of surprising features. First, for $0 \leq L < 1$, the density function is flat (if one overlooks the "bumps," which are fluctuations due to

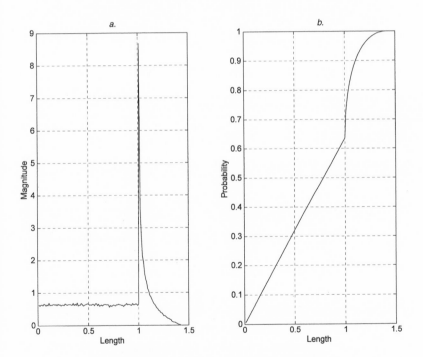

Figure 17.2  a. Density of $L$ from 100,000 Paths
b. Distribution of $L$ from 100,000 Paths

the finite size of the simulation sample; yes, even 100,000 is "finite."), something I don't think is obvious at all in the physical nature of random paths across a square. And second, there is a big jump in the density at $L = 1$ (as the following theoretical analysis will show, it is an infinite discontinuity in the density). The distribution function, as the integral of the density, is smoother and continuous everywhere, but it does suddenly change slope at $L = 1$ because of the density discontinuity. This shows up as an interesting "hump" when $L$ is between 1 and $\sqrt{2}$.

The key idea behind a theoretical analysis of this problem is the theorem of total probability, which says that for any given $\ell$,

$$P(L \leq \ell) = \int_0^1 P(L \leq \ell | x) \, f_X(x) \, dx, \; 0 \leq \ell \leq \sqrt{2}.$$

That is, to find the distribution function $P(L \leq \ell)$, we first condition it on a particular value of $X$ and then integrate the conditional distribution against the probability density of $X$, $f_X(x)$. Since $X$ is uniform from 0 to 1, we have $f_X(x) = 1$, so we need only to evaluate the quite simple-looking integral

$$P(L \leq \ell) = \int_0^1 P(L \leq \ell | x) \, dx.$$

As you'll soon see, it will make our calculations very much less complicated if we redefine how we measure the angle that a path makes with the $x$-axis. Rather than proceeding counterclockwise as in the original statement of the problem, I will measure $\theta$ in a clockwise way with respect to the horizontal axis, as shown in Figure 17.3. Also helpful is the realization that all we really need to consider is the interval of the angles $0 \leq \theta \leq \frac{\pi}{2}$ since, by symmetry, for every $x$ and $\theta$ in the $0 \leq \theta \leq \frac{\pi}{2}$ interval there is another

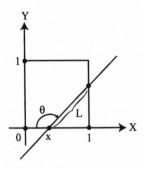

Figure 17.3 Measurement of $\theta$ with Respect to X-Axis

$x$ and $\theta$ in the $\frac{\pi}{2} \leq \theta \leq \pi$ interval that results in the same path length. That is, if we insist on doing the analysis for $0 \leq \theta \leq \pi$, then all we accomplish is the doubling of the number of paths that satisfy the condition $L \leq \ell$ along with the doubling of the total number of paths. Thus, the value of $P(L \leq \ell)$ will be unchanged whether $\theta$ is uniform from 0 to $\frac{\pi}{2}$ or uniform from 0 to $\pi$. So, from now on, I'll assume that $\theta$ is uniform from 0 to $\frac{\pi}{2}$.

To start, consider the case of $\ell \leq 1$. This means that we are interested only in paths that cut through the left vertical edge of the square, because to cut through the top edge requires a path length at least equal to 1. As illustrated in Figure 17.4, such paths occur only for angles $\theta$ in the interval $0 \leq \theta \leq \cos^{-1}\left(\frac{x}{\ell}\right)$. So, for $\ell \leq 1$,

$$P(L \leq \ell | x) = \frac{\cos^{-1}\left(\dfrac{x}{\ell}\right)}{\dfrac{\pi}{2}} = \frac{2}{\pi} \cos^{-1}\left(\frac{x}{\ell}\right)$$

and therefore

$$P(L \leq \ell) = \int_{0}^{\ell} \frac{2}{\pi} \cos^{-1}\left(\frac{x}{\ell}\right) dx, \ \ell \leq 1.$$

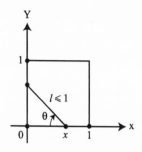

Figure 17.4  Paths that Cut Through Left Edge of Square

The upper limit on the integral is important to understand; the reason why $x$ cannot be greater than $\ell$ is simply
that if $x$ is greater than $\ell$, then any path through that $x$
that cuts the left vertical edge would have a length greater
than $\ell$.

Now, changing the dummy integration variable to $u =$
$x/\ell$, we have

$$P(L \leq \ell) = \frac{2}{\pi} \int\limits_0^1 \cos^{-1}(u)\, \ell\, du = \frac{2\ell}{\pi} \left[ u \cos^{-1}(u) - \sqrt{1 - u^2}\, \right]_0^1.$$

$$= \frac{2\ell}{\pi} \left[ \cos^{-1}(1) + 1 \right].$$

Since $\cos^{-1}(1) = 0$, we can neatly write that as

$$F_L(\ell) = P(L \leq \ell) = \frac{2\ell}{\pi},\ 0 \leq \ell \leq 1.$$

That is, $F_L(\ell)$ increases linearly over the interval $0 \leq \ell \leq$
1, a conclusion supported by the simulation result in the
right-hand plot of Figure 17.2.

Consider next the remaining case of $1 \leq \ell \leq \sqrt{2}$. In Figure 17.5 I've shown how, for any given $\ell \geq 1$, that there is

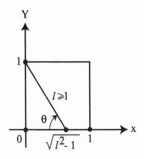

Figure 17.5  Specified Interval on the X-Axis

an interval on the $x$-axis for which all possible angles $0 \leq \theta \leq \frac{\pi}{2}$ lead to paths with lengths $L$ no longer than $\ell$. That is, for $0 \leq x \leq \sqrt{\ell^2 - 1}$ we have $P(L \leq \ell \mid x) = 1$. For the rest of the $x$-axis, however—that is for $\sqrt{\ell^2 - 1} \leq x \leq 1$, there are two distinct ways in which a path through a given $x$ can satisfy the condition $L \leq \ell$.

In Figure 17.6, I've shown the two paths with length $L = \ell \geq 1$ that are possible for a given $x \geq \sqrt{\ell^2 - 1}$. One is at angle $\theta_1$ and the other is at angle $\theta_2$, where

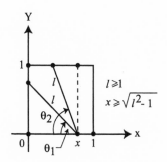

Figure 17.6  Two Possible Paths

$$\theta_1 = \cos^{-1}\left(\frac{x}{\ell}\right)$$

$$\theta_2 = \frac{\pi}{2} - \cos^{-1}\left(\frac{1}{\ell}\right).$$

For all angles in the interval $0 \leq \theta \leq \theta_1$, we have paths that cut the left vertical edge with lengths $L$ that satisfy the condition $L \leq \ell$, and for all angles in the interval $\theta_2 \leq \theta \leq \frac{\pi}{2}$ (an interval with width, of course, of $\cos^{-1}\left(\frac{1}{\ell}\right)$) we have paths that cut the top horizontal edge with lengths $L$ that also satisfy the condition $L \leq \ell$. Our problem now is to calculate the integral

$$P(L \leq \ell) = \int_0^1 P(L \leq \ell \mid x)\, dx, \ 1 \leq \ell \leq \sqrt{2}.$$

From earlier in this analysis, we have $P(L \leq \ell \mid x) = 1$ for $0 \leq x \leq \sqrt{\ell^2 - 1}$; now, we can see that for $x \geq \sqrt{\ell^2 - 1}$,

$$P(L \leq \ell \mid x) = \frac{\cos^{-1}\left(\frac{x}{\ell}\right) + \cos^{-1}\left(\frac{1}{\ell}\right)}{\frac{\pi}{2}}$$

$$= \frac{2}{\pi}\left[\cos^{-1}\left(\frac{x}{\ell}\right) + \cos^{-1}\left(\frac{1}{\ell}\right)\right], \ \sqrt{\ell^2 - 1} \leq x \leq 1.$$

So,

$$P(L \leq \ell) = \int_0^{\sqrt{\ell^2 - 1}} dx + \frac{2}{\pi}\int_{\sqrt{\ell^2-1}}^1 \left[\cos^{-1}\left(\frac{x}{\ell}\right) + \cos^{-1}\left(\frac{1}{\ell}\right)\right] dx$$

$$= \sqrt{\ell^2 - 1} + \frac{2}{\pi}\cos^{-1}\left(\frac{1}{\ell}\right)\left[1 - \sqrt{\ell^2 - 1}\right]$$

$$+ \frac{2}{\pi}\int_{\sqrt{\ell^2-1}}^1 \cos^{-1}\left(\frac{x}{\ell}\right) dx.$$

Or, changing the dummy integration variable to $u = x/\ell$ in the integral,

$$P(L \leq \ell) = \sqrt{\ell^2 - 1} + \frac{2}{\pi} \left\{ \cos^{-1}\left(\frac{1}{\ell}\right)[1 - \sqrt{\ell^2 - 1}] + \right.$$

$$\left. \ell \int_{\frac{\sqrt{\ell^2-1}}{\ell}}^{1/\ell} \cos^{-1}(u) \, du \right\}.$$

For the integral, we have

$$\int_{\frac{\sqrt{\ell^2-1}}{\ell}}^{1/\ell} \cos^{-1}(u) \, du = \left[ u \cos^{-1}(u) - \sqrt{1 - u^2} \right]_{\frac{\sqrt{\ell^2-1}}{\ell}}^{1/\ell}$$

$$= \frac{1}{\ell} \cos^{-1}\left(\frac{1}{\ell}\right) - \sqrt{1 - \frac{1}{\ell^2}} - \frac{\sqrt{\ell^2 - 1}}{\ell}$$

$$\cos^{-1}\left(\frac{\sqrt{\ell^2 - 1}}{\ell}\right) + \frac{1}{\ell}.$$

Or,

$$\int_{\frac{\sqrt{\ell^2-1}}{\ell}}^{1/\ell} \cos^{-1}(u) \, du = \cos^{-1}\left(\frac{1}{\ell}\right) - \sqrt{\ell^2 - 1}$$

$$- \sqrt{\ell^2 - 1} \, \cos^{-1}\left(\frac{\sqrt{\ell^2 - 1}}{\ell}\right) + 1.$$

Thus, for $1 \leq \ell \leq \sqrt{2}$, we have

$$P(L \leq \ell) = \sqrt{\ell^2 - 1} + \frac{2}{\pi} \left\{ \cos^{-1}\left(\frac{1}{\ell}\right) - \cos^{-1}\left(\frac{1}{\ell}\right) \sqrt{\ell^2 - 1} \right.$$

$$+ \cos^{-1}\left(\frac{1}{\ell}\right)$$

$$- \sqrt{\ell^2 - 1} - \sqrt{\ell^2 - 1}$$

$$\left. \cos^{-1}\left(\frac{\sqrt{\ell^2 - 1}}{\ell}\right) + 1 \right\}$$

$$= \sqrt{\ell^2 - 1} + \frac{2}{\pi} \left\{ 1 + 2 \cos^{-1}\left(\frac{1}{\ell}\right) \right.$$

$$\left. - \sqrt{\ell^2 - 1} \left[ \cos^{-1}\left(\frac{1}{\ell}\right) + \cos^{-1}\left(\frac{\sqrt{\ell^2 - 1}}{\ell}\right) + 1 \right] \right\}.$$

Since $\cos^{-1}\left(\dfrac{1}{\ell}\right) + \cos^{-1}\left(\dfrac{\sqrt{\ell^2 - 1}}{\ell}\right) = \dfrac{\pi}{2}$, this reduces to

$$P(L \leq \ell) = \frac{2}{\pi} \left[ 1 + 2 \cos^{-1}\left(\frac{1}{\ell}\right) - \sqrt{\ell^2 - 1} \right].$$

So, in summary,

$$F_L(\ell) = \begin{cases} \dfrac{2\ell}{\pi}, & 0 \leq \ell \leq 1 \\[2ex] \dfrac{2}{\pi} \left[ 1 + 2 \cos^{-1}\left(\dfrac{1}{\ell}\right) - \sqrt{\ell^2 - 1} \right], & 1 \leq \ell \leq \sqrt{2}. \end{cases}$$

Notice that the two expressions for $F_L(\ell)$ are equal at $\ell = 1$, i.e., the distribution function is continuous, and also that $F_L(0) = 0$ and $F_L(\sqrt{2}) = 1$, all of which are the properties of any distribution function.

Figure 17.7 is a plot of $F_L(\ell)$, produced by the program **theory.m** (Program 29), over the entire interval $0 \leq \ell \leq \sqrt{2}$; notice that it looks just like the result of the simulation plot of the distribution in Figure 17.2. The pdf for $L$ is found by differentiating $F_L(\ell)$, so

$$f_L(\ell) = \begin{cases} \dfrac{2}{\pi}, & 0 \leq \ell \leq 1 \\[2ex] \dfrac{2}{\pi} \cdot \dfrac{2 - \ell^2}{\ell \sqrt{\ell^2 - 1}}, & 1 < \ell \leq \sqrt{2}. \end{cases}$$

Thus, the pdf has an infinite discontinuity at $\ell = 1$. Notice, too, that $f_L(\sqrt{2}) = 0$, just as indicated by the simulation curve in the left-hand plot of Figure 17.2.

154    The Solutions

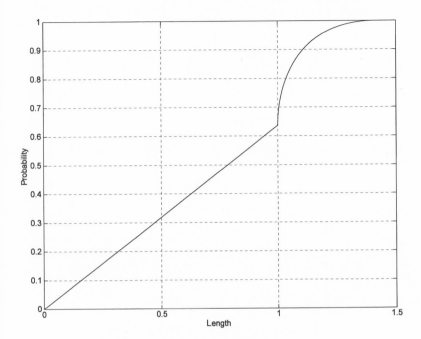

Figure 17.7 Theoretical Distribution Function

## 18. Two Flies Stuck on a Piece of Flypaper—How Far Apart?

To begin, I'll solve a preliminary problem. Let $X$ and $Y$ be uniform from 0 to 1, and independent. How does $Z = (X-Y)^2$ behave? Clearly, $0 \le Z \le 1$. Also, we have the distribution of $Z$ as

$$F_Z(z) = P(Z \le z) = P[(X-Y)^2 \le z] = P[-\sqrt{z} \le X - Y \le \sqrt{z}]$$

$$= P[\sqrt{z} \ge Y - X \ge -\sqrt{z}] = P[X + \sqrt{z} \ge Y \ge X - \sqrt{z}].$$

This last probability is the probability of the shaded region shown in Figure 18.1. From geometric probability theory, it

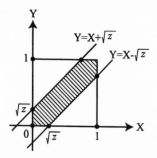

Figure 18.1  Probability of Shaded Region

is just the ratio of the area of the shaded region to the area of the entire sample space (unity), so

$$F_Z(z) = 1 - (1 - \sqrt{z})^2, \, 0 \le z \le 1.$$

Notice that $F_Z(0) = 0$ and $F_Z(1) = 1$, just as we would expect a distribution function to behave. Finally, the density function of $Z$ is

$$f_Z(z) = \frac{d}{dz} F_Z(z) = \frac{1 - \sqrt{z}}{\sqrt{z}}, \, 0 \le z \le 1.$$

Now, returning to our original problem, define $X = (X_1 - X_2)^2$ and $Y = (Y_1 - Y_2)^2$. The preliminary result above immediately tells us that the density functions for $X$ and $Y$ are

$$f_X(x) = \frac{1 - \sqrt{x}}{\sqrt{x}}, \, 0 \le x \le 1$$

$$f_Y(y) = \frac{1 - \sqrt{y}}{\sqrt{y}}, \, 0 \le y \le 1.$$

Our problem is to calculate $P(X + Y \ge t) = P(Y \ge t - X)$ and, as Figure 18.2 shows, there are two cases to consider. That is, the geometry of the problem is different for $0 \le t$

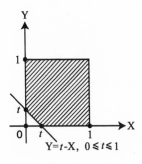

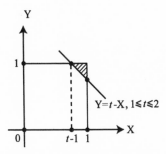

Figure 18.2   Geometry Is Different for $0 \leq t \leq 1$ and $1 \leq t \leq 2$

$\leq 1$ and for $1 \leq t \leq 2$. In both cases, however, $P(Y > t - X)$ is the probability of the shaded region. That probability, in turn, is simply the integral of the joint probability density function of $X$ and $Y$ over the shaded region. Because $X$ and $Y$ are independent, the joint pdf is the product of the individual (or marginal) pdfs, so

$$f_{X,Y}(x,\, y) = \frac{1 - \sqrt{x}}{\sqrt{x}} \cdot \frac{1 - \sqrt{y}}{\sqrt{y}},\ 0 \leq x,\, y \leq 1.$$

So, for the case of $0 \leq t \leq 1$, we have

$$P(X + Y \geq t) = 1 - \int_{0}^{t} \left[ \int_{0}^{t-x} \frac{1 - \sqrt{x}}{\sqrt{x}} \cdot \frac{1 - \sqrt{y}}{\sqrt{y}}\, dy \right] dx$$

$$= 1 - \int_0^t \frac{1-\sqrt{x}}{\sqrt{x}} \left[ \int_0^{t-x} \frac{1-\sqrt{y}}{\sqrt{y}} \, dy \right] dx,$$

where I have subtracted the probability of the unshaded region (where $X + Y < t$) from 1. For the $y$ integral, we have

$$\int_0^{t-x} \frac{1-\sqrt{y}}{\sqrt{y}} \, dy = \int_0^{t-x} \left( \frac{1}{\sqrt{y}} - 1 \right) dy = \int_0^{t-x} \frac{dy}{\sqrt{y}} - (t-x)$$

$$= x - t + \int_0^{t-x} y^{-\frac{1}{2}} \, dy = x - t + \left( 2y^{\frac{1}{2}} \Big|_0^{t-x} \right)$$

$$= x - t + 2\sqrt{t-x}.$$

Thus, for $0 \le t \le 1$, we have

$$P(X + Y \ge t) = 1 - \int_0^t \frac{1-\sqrt{x}}{\sqrt{x}} \left( x - t + 2\sqrt{t-x} \right) dx$$

$$= 1 - \int_0^t \frac{x - t + 2\sqrt{t-x} - x\sqrt{x} + t\sqrt{x} - 2\sqrt{x}\sqrt{t-x}}{\sqrt{x}} \, dx$$

$$= 1 - \int_0^t \left( \sqrt{x} - \frac{t}{\sqrt{x}} + 2\sqrt{\frac{t}{x} - 1} - x + t \right.$$

$$\left. - 2\sqrt{t-x} \right) dx$$

$$= 1 - \left[ \frac{2}{3} x^{\frac{3}{2}} - 2tx^{\frac{1}{2}} - \frac{1}{2} x^2 + tx + \frac{4}{3} (t-x)^{\frac{3}{2}} \right]_0^t$$

$$- 2\int_0^t \sqrt{\frac{t}{x} - 1} \, dx$$

$$= 1 - \left( \frac{2}{3} t^{\frac{3}{2}} - 2t^{\frac{3}{2}} - \frac{1}{2} t^2 + t^2 - \frac{4}{3} t^{\frac{3}{2}} \right)$$

$$- 2\int_0^t \sqrt{\frac{t}{x} - 1} \, dx$$

$$= 1 + \frac{8}{3} t^{\frac{3}{2}} - \frac{1}{2} t^2 - 2 \int_0^t \sqrt{\frac{t}{x} - 1} \, dx.$$

For the last term (the integral), change the dummy integration variable to

$$u = \sqrt{\frac{t}{x} - 1} = \left( \frac{t}{x} - 1 \right)^{1/2},$$

so
$$u^2 = \frac{t}{x} - 1 \quad \text{or,} \quad x = \frac{t}{u^2 + 1}.$$

Then,

$$\frac{du}{dx} = \frac{1}{2} \left( \frac{t}{x} - 1 \right)^{-\frac{1}{2}} \left( -\frac{t}{x^2} \right) = -\frac{t}{2x^2 (\frac{t}{x} - 1)^{\frac{1}{2}}} = -\frac{t}{2 \frac{t^2}{(u^2 + 1)^2} u}$$

or,

$$\frac{du}{dx} = -\frac{(u^2 + 1)^2}{2ut}$$

or,

$$dx = -\frac{2ut}{(u^2 + 1)^2} \, du.$$

So, at last,

$$\int_0^t \sqrt{\frac{t}{x} - 1} \, dx = \int_0^\infty u \left[ -\frac{2ut}{(u^2 + 1)^2} \, du \right] = 2t \int_0^\infty \frac{u^2}{(u^2 + 1)^2} \, du.$$

From integral tables, we find that

$$2t \int_0^\infty \frac{u^2}{(u^2 + 1)^2} \, du = 2t \left( -\frac{1}{2} \frac{u}{u^2 + 1} + \frac{1}{2} \tan^{-1}(u) \right) \Big|_0^\infty$$

$$= 2t \cdot \frac{1}{2} \tan^{-1}(\infty) = \frac{1}{2} \pi t.$$

Our answer then, is

$$P(X + Y \geq t) = 1 + \frac{8}{3} t^{\frac{3}{2}} - \frac{1}{2} t^2 - \pi t, \ 0 \leq t \leq 1.$$

In particular, for $t = 1$, we have

$$P(X + Y \geq 1) = 1 + \frac{8}{3} - \frac{1}{2} - \pi = \frac{19}{6} - \pi = 0.025074,$$

which is "pretty small."

This calculation is all we need to do to answer the original question, but as a check we could also do the integral for the case of $1 \leq t \leq 2$ to see if it is consistent with our first calculation. I won't go through the details of that second integral here, but if you are looking for an interesting exercise in integration, here is the answer for $1 \leq t \leq 2$:

$$P(X + Y \geq t) = \int_{t-1}^{1} \left[ \int_{t-x}^{1} \frac{1 - \sqrt{x}}{\sqrt{x}} \cdot \frac{1 - \sqrt{y}}{\sqrt{y}} \, dy \right] dx$$

$$= \frac{2}{3} + \frac{1}{2} t^2 + 2t - \frac{2}{3} (t-1)^{\frac{3}{2}} - 2t \, (t-1)^{\frac{1}{2}}$$

$$- 2(t-1)^{\frac{1}{2}}$$

$$- 2t \left[ \tan^{-1} \left( \frac{1}{\sqrt{t-1}} \right) - \tan^{-1} \left( \sqrt{t-1} \right) \right], \ 1 \leq t \leq 2.$$

What an expression, just to account for a probability of 0.025074! Indeed, notice that for $t = 1$ this expression does reduce to $\frac{19}{6} - \pi$, and that for $t = 2$ it reduces to 0 (just as it should).

To check this theoretical result experimentally, the MATLAB program **flysquare.m** (Program 30) simulates the landing of 100,000 pairs of flies on the unit square, and keeps track of how many pairs have a separation distance squared greater than one. The result, from one run, was the estimate $P[(X_1 - X_2)^2 + (Y_1 - Y_2)^2 > 1] = 0.02524$, which is fairly close to the theoretical result. Finally, the program **flycircle1.m** (Program 31) simulates the random landing of 100,000 pairs of flies on a circle with unit area (see the program itself for what "random" means in this case), with the result of one run being $P[(X_1 - X_2)^2 + (Y_1 - Y_2)^2 > 1] = 0.00466$, considerably less than the estimate for the unit square.

## 19. The Blind Spider and the Fly

The program **spider.m** (Program 32) performs a random walk on any two-dimensional web that the user describes. To describe a web, I've used the following simple data structure: For each vertex, define a row vector whose first element is the number of vertices to which that vertex connects, with that number of next elements in the row vector listing the specific vertex numbers of those vertices. All of these row vectors, together, form the matrix in **spider.m** called **web**, where unused entries are set equal to zero—which explains why all the entries for vertex 2 are equal to zero, as the walk stops at that vertex. Thus, the row vectors for the web of Figure 19.1 are:

| Vertex | | | | | |
|---|---|---|---|---|---|
| 1 | 3 | 4 | 7 | 8 | 0 |
| 2 | 0 | 0 | 0 | 0 | 0 |
| 3 | 3 | 2 | 5 | 8 | 0 |
| 4 | 1 | 1 | 0 | 0 | 0 |
| 5 | 2 | 3 | 6 | 0 | 0 |
| 6 | 2 | 5 | 7 | 0 | 0 |
| 7 | 3 | 1 | 6 | 8 | 0 |
| 8 | 4 | 1 | 3 | 7 | 9 |
| 9 | 2 | 2 | 8 | 0 | 0 |

For example, vertex 5 connects to two other vertices (numbers 3 and 6), while vertex 8 connects to four other vertices (numbers 1, 3, 7, and 9). For ease of programming, the web matrix is described by coding it in the form of column vectors (the t1 to t5 vectors in **spider.m**) and then forming the web from the column vectors in the "web =" statement. With this web structure, I think the rest of the MATLAB code is self-explanatory.

When I ran **spider.m** several times, each for 10,000 walks, it consistently gave an average value of about 15.4 steps; the plot of the distribution of walk durations shown in Figure 19.3 is typical. To see if the code gives the correct answer to a problem for which we already know the

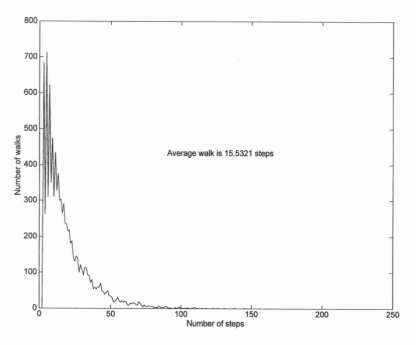

Figvre 19.3   Distribution of Walk Durations on the Web (10,000 walks)

answer, I tried the one-dimensional web of length $n = 3$, where the vertices are numbered as follows:

spider                          fly

•————•————•————•

1          3          4          2

The web matrix is

| Vertex | | | |
|---|---|---|---|
| 1 | 1 | 3 | 0 |
| 2 | 0 | 0 | 0 |
| 3 | 2 | 1 | 4 |
| 4 | 2 | 2 | 3 |

When I entered this into **spider.m** (as three column vectors) it produced, in two runs, the average values of 8.7898 and 8.9145, which are very close to the theoretical answer of $n^2 = 3^2 = 9$.

Finally, although it wasn't part of your assignment, consider again the Brownian motion random walk of Figure 19.2. In the MATLAB program **brownian.m**, each new step motion of the particle was simulated by selecting two independent random numbers ($\Delta x$ and $\Delta y$) uniformly from the interval $-1$ to $1$. Let's write $X_1, X_2, \ldots, X_n$ and $Y_1, Y_2, \ldots, Y_n$ to denote these deltas (i.e., the changes) in the $x$ and $y$ coordinates of the particle, respectively, at the end of the first, second, through $n$th steps. The radial distance of the particle from the origin, at the end of the $n$th step, is then $R_n$, where

$$R_n^2 = (X_1 + X_2 + \ldots + X_n)^2 + (Y_1 + Y_2 + \ldots + Y_n)^2$$

$$= \sum_{i=1}^{n} (X_i^2 + Y_i^2) + \sum_{\substack{i=1 \\ i \neq j}}^{n} \sum_{j=1}^{n} (X_i X_j + Y_i Y_j).$$

The average or expected value of $R_n^2$ is, then,

$$E(R_n^2) = \sum_{i=1}^{n} \left[ E(X_i^2) + E(Y_i)^2 \right] + \sum_{\substack{i=1 \\ i \neq j}}^{n} \sum_{j=1}^{n} \left[ E(X_i X_j) + E(Y_i Y_j) \right].$$

Since the $X_i$ and $Y_i$ are all independent of one another, and since all are uniform from $-1$ to $1$, then when $i \neq j$,

$$E(X_i X_k) = E(Y_i Y_j) = E(X_i)E(X_j) = E(Y_i)E(Y_j) = 0.$$

And when $i = j$,

$$E(X_i^2) = E(Y_j^2) = \int_{-1}^{1} x^2 \frac{1}{2} \, dx = \frac{1}{2}\left( \frac{1}{3} x^3 \Big|_{-1}^{1} \right) = \frac{1}{3}.$$

Thus, $E(R_n^2) = \frac{2}{3} n$. Or, taking the square root to get the root of the mean of the square distance, or the "rms" radial distance, $\sqrt{E(R_n^2)} = 0.8165 \sqrt{n}$. This says, for example, that a walk with a thousand steps will, compared to a walk with a hundred steps, increase the particle's distance (on average) from the origin by only a factor of about three.

As a final comment, the one-dimensional random walk with absorbing barriers at both $x = 0$ and $x = n$ is famous as the "Gambler's Ruin," which dates from a problem posed in 1654 by Blaise Pascal (1623–1662) to his fellow Frenchman Pierre de Fermat (1601–1665). It is so called because it can be thought of as describing the evolution of a gambler's wealth as he places a succession of \$1 bets. He takes one step to the right if he wins a bet (with probability $p$) and one step to the left if he loses (with probability $q = 1 - p$). The value of $n$ is the total amount of money available—the gambler's plus the opponent's—so if $x = n$ the gambler has won it all and if $x = 0$ he has lost it all, i.e., is ruined. If $\omega_k$ is the probability that the gambler wins it all if he starts with \$$k$—and thus the opponent starts with \$$(n-k)$—then $\omega_k = p\omega_{k+1} + q\omega_{k-1}$ where $\omega_0 = 0$ and $\omega_n = 1$, since if the gambler wins the first bet he then has \$$(k+1)$ and he has \$$(k-1)$ if he loses. This is a second-order recurrence with the solution

$$\omega_k = \frac{1 - \left(\dfrac{q}{p}\right)^k}{1 - \left(\dfrac{q}{p}\right)^n}, \; p \neq \frac{1}{2}.$$

Even if you can't solve the recurrence, you can check this solution's correctness by substitution back into the recurrence. The recurrence solution is credited to James Bernoulli (1654–1705), and it appears in his *Ars Conjectandi* (1713).

## 20. Reliably Unreliable

1. Using the hint given in the original problem statement, we have

$$f_X(x)\Delta x = \left[1 - e^{-a(x+\Delta x)}\right] - \left[1 - e^{-ax}\right]$$
$$= e^{-ax} - e^{-a(x+\Delta x)}$$

so

$$f_X(x) = \frac{e^{-ax} - e^{-a(x+\Delta x)}}{\Delta x} = -\frac{e^{-a(x+\Delta x)} - e^{-ax}}{\Delta x}.$$

But if we define $g(x) = e^{-ax}$, then this is simply

$$f_X(x) = -\frac{g(x+\Delta x) - g(x)}{\Delta x}$$

which, in the limit as $\Delta x \to 0$, is the definition of the negative derivative of $g(x)$. That is,

$$f_X(x) = -\frac{dg}{dx} = ae^{-ax}, \; x \geq 0.$$

From this we have the distribution function, $F_X(x)$, as

$$F_X(x) = P(X \leq x) = \int_0^x f_X(u) \; du$$

$$= \int_0^x ae^{-au} \; du = a\left(-\frac{1}{a}e^{-au}\right)\Big|_0^x$$

$$= 1 - e^{-ax}, \; x \geq 0.$$

2.

$$E(X) = \int_0^\infty x f_X(x) \; dx = \int_0^\infty axe^{-ax} \; dx$$

$$= a\left[-\frac{e^{-ax}}{a}\left(\frac{1}{a}+x\right)\right]\Big|_0^\infty = \frac{1}{a}.$$

That is, the physical significance of the constant $a$ is that it is the reciprocal of the average lifetime of a system.

3. To find the average lifetime of a SYSTEM, made of three identical copies of a system that are connected in parallel in such a way that the SYSTEM fails when the second system fails, write

$P(Z \leq z) = F_Z(z) = P(\text{SYSTEM fails by time } z)$
$= 1 - P(Z > z) = 1 - P(\text{SYSTEM still working at time } z)$
$= 1 - P(\text{all three systems or any two systems still working at time } z).$

Let's denote the random lifetimes of the three systems by

$X_1$, $X_2$, and $X_3$; i.e., by $X_i$ with $i = 1, 2, 3$. Then the above says

$$F_Z(z) = 1 - \Big[ P(\text{all three } X_i > z) + P(\text{any two } X_i > z) \Big].$$

Next, because we are assuming the $X_i$ are independent, we have

$$F_Z(z) = 1 - \begin{bmatrix} P(X_1 > z)\, P(X_2 > z)\, P(X_3 > z) \\ + P(X_1 > z)\, P(X_2 > z)\, P(X_3 < z) \\ + P(X_1 > z)\, P(X_2 < z)\, P(X_3 > z) \\ + P(X_1 < z)\, P(X_2 > z)\, P(X_3 > z) \end{bmatrix}.$$

Since the $X_i$ are statistically identical, we can drop the subscripts; thus,

$$F_Z(z) = 1 - \{ [1 - F_X(z)]^3 + 3F_X(z)\, [1 - F_X(z)]^2 \}$$

or,  $\quad F_Z(z) = 3\, F_X^2(z) - 2\, F_X^3(z).$

Notice that $F_Z(0) = 3F_X^2(0) - 2F_X^3(0) = 0$ and that $F_Z(\infty) = 3F_X^2(\infty) - 2F_X^3(\infty) = 3 - 2 = 1$, which are indeed the proper endpoint values for the distribution function $F_Z(z)$.

From $F_Z(z)$ we can find $f_Z(z)$ by differentiation:

$$f_Z(z) = \frac{d}{dz}\, F_Z(z) = 6F_X(z)\, f_X(z) - 6F_X^2(z)\, f_X(z)$$

$$= 6F_X(z)\, f_X(z)[1 - F_X(z)].$$

Inserting the earlier expressions for $F_X(z)$ and $f_X(z)$, we arrive at

$$f_Z(z) = 6a\Big[ e^{-2az} - e^{-3az} \Big], \quad z \geq 0.$$

4. Thus, the average value of $Z$ is calculated as

$$E(Z) = \int_0^\infty z f_Z(z)\, dz = 6a \left[ \int_0^\infty z e^{-2az}\, dz - \int_0^\infty z e^{-3az}\, dz \right]$$

$$= 6a \left[ \left( -\frac{e^{-2az}}{2a} \left\{ \frac{1}{2a} + z \right\} \Big|_0^\infty \right) - \left( -\frac{e^{-3az}}{3a} \left\{ \frac{1}{3a} + z \right\} \Big|_0^\infty \right) \right].$$

$$= 6a \left[ \frac{1}{4a^2} - \frac{1}{9a^2} \right] = \frac{6}{a} \left( \frac{1}{4} - \frac{1}{9} \right) = \frac{5}{6} \cdot \frac{1}{a}.$$

That is, $E(Z) = \frac{5}{6} E(X)$, i.e., the average lifetime of the parallel redundant SYSTEM is less than the average lifetime of an individual system. Isn't this a paradox? After all, why do engineers actually make parallel redundant SYSTEMs when they won't, on average, last as long as even one of the component systems?

One answer might be that perhaps the average lifetime is the wrong measure to use; why not, for example, use the *median* lifetime, the time at which half the parallel SYSTEMs have failed? For the individual systems, the median lifetime is $x_0$, where

$$P(X \le x_0) = \frac{1}{2} = F_X(x_0) = 1 - e^{-ax_0}.$$

This is quickly solved to give

$$x_0 = \frac{1}{a} \ln(2) = E(X)\ln(2) = 0.693\ E(X).$$

The median lifetime for the SYSTEM is $z_0$, where

$$P(Z \le z_0) = \frac{1}{2} = F_Z(z_0) = 3F_X^2(z_0) - 2F_X^3(z_0).$$

With just a bit of easy algebra, this becomes

$$e^{-3az_0} - \frac{3}{2}^{-2az_0} + \frac{1}{4} = 0.$$

This may look awful to solve, but consider the following: Let $x = e^{-az_0}$. Then, we have

$$x^3 - \frac{3}{2}x^2 + \frac{1}{4} = 0,$$

for which you can perhaps see that $x = \frac{1}{2}$ is a solution. Dividing through the cubic by $x - \frac{1}{2}$, we reduce the equation to the quadratic $x^2 - x - \frac{1}{2} = 0$, which has the additional two real solutions $x = \frac{1}{2} + \frac{1}{2}\sqrt{3}$ and $x = \frac{1}{2} - \frac{1}{2}\sqrt{3}$. Since our solution for $z_0$ is

$$z_0 = -\frac{1}{a}\ln(x) = -E(X)\ln(x),$$

we can eliminate the root $x = \frac{1}{2} - \frac{1}{2}\sqrt{3}$, since it is negative and would give an imaginary median lifetime. We can also eliminate the root $x = \frac{1}{2} + \frac{1}{2}\sqrt{3}$, since it is greater than 1 and would give a negative median lifetime (perhaps not quite so bad as an imaginary result, but bad enough). Only the $x = \frac{1}{2}$ root gives a physically reasonable answer, i.e.,

$$z_0 = E(X)\ln(2).$$

However, since this is exactly the same median lifetime we got for an individual system, the added complexity of the parallel redundant SYSTEM still seems not to have bought us anything, and the paradox seemingly remains.

Not to keep you in suspense any longer, the answer is revealed when you plot the distribution functions for $X$

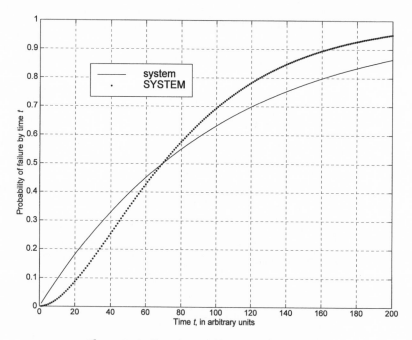

Figure 20.1  Distribution Functions for X and Z

and Z on the same set of axes, as does the MATLAB program **paradox.m** (Program 33) in Figure 20.1, for the
illustrative value of $E(X) = 100$ (in arbitrary units of
time). These curves show that, indeed, X (an individual
system) and Z (the parallel redundant SYSTEM) do have
equal median lifetimes (the curves cross at $\frac{1}{2}$ on the vertical axis). But, they also show that for all times less than
the median lifetime, the SYSTEM is less likely to fail
than is a single system. So, as long as we use the
SYSTEM for missions lasting no longer than the median
lifetime, then the SYSTEM is more reliable than is a
system.

# 21. When Theory Fails, There Is Always the Computer

Applying the CPM algorithm to Project A, described in Figure 21.4, results in the analysis shown in Figure 21.6. That is, the total time required for the project is 30. This is confirmed when the following data statements are entered into **cpm.m** (with **ntasks** = 11;),

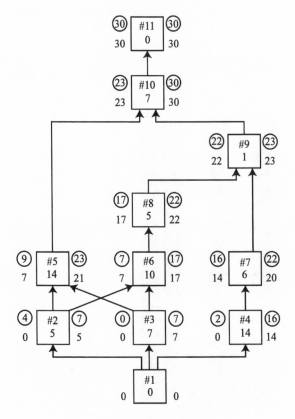

Figure 21.6   Analysis of Project A

data $(1,:) = [0, 0, 0, 0]$;
data $(2,:) = [1, 1, 5, 0]$;
data $(3,:) = [1, 1, 7, 0]$;
data $(4,:) = [1, 1, 14, 0]$;
data $(5,:) = [2, 2, 3, 14]$;
data $(6,:) = [2, 2, 3, 10]$;
data $(7,:) = [1, 4, 6, 0]$;
data $(8,:) = [1, 6, 5, 0]$;
data $(9,:) = [2, 7, 8, 1]$;
data $(10,:) = [2, 5, 9, 7]$;
data $(11,:) = [1, 10, 0, 0]$;

which produces the following output:
Total time required for project = 30

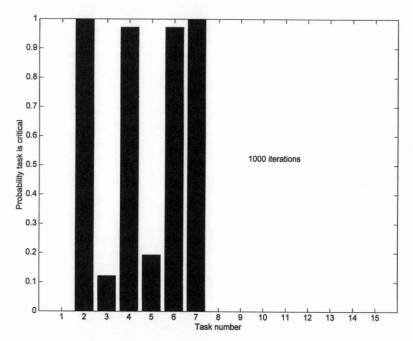

Figure 21.7   Critical Task Likelihood for Each Task of Fig. 21.5

| Task | Early start time | Slack |
|------|------------------|-------|
| 2 | 0 | 2 |
| 3 | 0 | 0 |
| 4 | 0 | 2 |
| 5 | 7 | 2 |
| 6 | 7 | 0 |
| 7 | 14 | 2 |
| 8 | 17 | 0 |
| 9 | 22 | 0 |
| 10 | 23 | 0 |

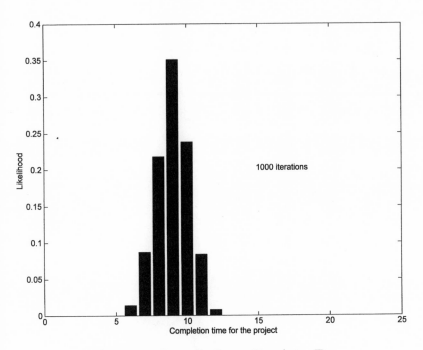

Figure 21.8 Range for Project Completion Time

The program **pert.m** (Program 34) is **cpm.m** modified to run the CPM algorithm using minimum/maximum task completion times. The data statements describe Project B of Figure 21.5, and the program output is in the form of two figures. The project completion times in Figure 21.7 show that tasks 2 and 7 are, at least statistically, always critical. Figure 21.8 looks bell-shaped, suggesting that it is a normally distributed random variable with an average value of 9. Indeed, the project completion time varied from 6 to 12, although the extreme times had a low probability of occurring.

# Random Number Generators

> Any one who considers arithmetical methods of producing random digits is, of course, in a state of sin. For ... there is no such thing as a random number—there are only methods to produce random numbers, and a strict arithmetic procedure of course is not such a method.
>
> —John von Neumann, in a 1951 paper in which the use of the plural *numbers* is crucial in understanding the distinction he is making.

The very idea of a deterministic machine like a computer creating random numbers seems to be an oxymoron. Random numbers are, well, *random*, while a computer is supposed to be an utterly predictable gadget. Indeed, no one has ever seen an advertisement saying something like

BUY OUR NEW SUPER MEGABLASTER COMPUTER!
YOU NEVER KNOW *WHAT* IT WILL DO!
DON'T BUY A DULL, BORING, PREDICTABLE COMPUTER;
BUY OURS AND BE SURPRISED EVERY TIME YOU TURN IT ON!

Would you buy such a machine? Probably not. And in fact, modern computers actually generate what are more precisely called *pseudo*-random numbers; that is, numbers that are produced by a deterministic algorithm but which, nonetheless, pass a series of statistical tests (devised by some of the world's best mathematicians) that say the numbers at least "appear" to be random. The subject of appearance is one that can quickly sink into a philosophical swamp, however, so let's just accept the fact that there are algorithms that pass these tests and devote the rest of this chapter to seeing how they work.

It is amusing to note first, however, that at least one of the better writers of 1950s science fiction, Theodore Sturgeon, completely missed the mark on number generators in one of his tales of the future. "The Pod in the Barrier" (Galaxy, September 1957) uses the classic science fiction technique of first presenting the reader with an apparently unsolvable problem, and then having a story character find a surprise (but scientifically rational) solution. In this particular story, the problem is how to defeat a seemingly impregnable missile defense system created by a super-advanced alien civilization. As one of the preliminary red herrings in his story, Sturgeon has two characters deliver the following exchange after one has told the other (and us) that the missiles are directed by radio signals such that "the frequency and amplitude of the control pulses make like purest noise—they're genuinely random.

" 'What do you mean, random? You can't control anything with random noise.'

"The captain thumbed over his shoulder [at the galactic location of the alien missile builders]. 'They can. There's a synchronous generator in the missiles that reproduces the *same random noise* [Sturgeon's emphasis], peak by pulse. ... I don't know *how* they do it. They just do.'

"England put his head down almost to the table. 'The same random,' he whispered from the very edge of sanity."

Despite those words of shock, by 1957 the mere mathematicians of earth had known for a decade how to do what Sturgeon seems to have believed only a super-advanced alien civilization could do. So, what *did* earth's mathematicians know then about the generation of random numbes?

In the very early days of computers (1948), the Princeton University mathematical genius John von Neumann (1903–1957) suggested the famous, but no longer used, middle square method for generating pseudo-random numbers. It is a method that is interesting for historical reasons, is very easy to understand, and has a certain immediate but ultimately superficial appeal (pointing out its problems is really the only reason for discussing it today). Von Neumann's interest in random number generation was sparked by a 1946 discussion he had with Stanislaw Ulam (mentioned in the preface), on the use of the Monte Carlo method as a way to answer many of the difficult mathematical questions related to the development of the atomic bomb.

The Monte Carlo method depends on access to a copious source of random numbers, and if one is going to do Monte Carlo on an electronic computer, then that source needs to be compatibly fast (looking numbers up in a table by hand[1] simply won't do). To allow initial program debugging to be done, the random number source should be repeatable, which eliminates using a natural physical process such as radioactive decay (e.g., counting the clicks on a Geiger counter over some interval of time, which would be far too slow in any case). It was von Neumann, then, who first recognized the need for an algorithmic procedure to quickly generate random numbers on demand, "en masse." Hence, the middle square method.

The algorithm is simplicity itself. Suppose we want to generate four-digit integers at random, i.e., numbers from 0000 to 9999. The middle-square method starts with an initial four-digit integer called the seed (say, 2437), squares it to get an eight-digit integer (05938969), and takes the middle four digits (9389) as the next random number. That number is then squared (88153321) and the middle four digits (1533) become the third random number, and so on. Each number in the sequence therefore uniquely determines the next number (but notice that the converse is not true).

Obviously, any sequence of four-digit numbers generated by this algorithm must eventually begin to repeat after a finite time; the number of numbers before repetition occurs is called the *period* of the sequence. The middle square method can easily degenerate into sequences that have very short periods; indeed, if 0 ever occurs, that number obviously reproduces itself immediately and the period is 1! For this reason alone, the search was soon on for other ways to generate random-appearing numbers.

In 1948 the *linear congruential generator* (l.c.g.) was proposed; essentially, all the random generators in use up to modern times are based on the l.c.g. algorithm. It was discussed first by the University of California number theorist D. H. Lehmer (1905–1991), a pure mathematician who also had an intense interest in the practical engineering details of computer hardware. The l.c.g. algorithm, as does the middle square method, starts with an initial seed integer (call it $X_0$) and generates all the subsequent random integers in its sequence from it according to the following deterministic rule:

$$X_{n+1} = (aX_n + c)\bmod m$$

where $a$, $c$, and $m$ are certain carefully selected integer constants. The notation "mod $m$" (or modulus $m$) means that $aX_n + c$ is divided by $m$ and $X_{n+1}$ is set equal to the remainder; i.e., $X_{n+1}$ takes on values from the integers 0, 1,

2, ... , $m-1$. For example, the choices of $a=13$, $c=0$ (which reduces the l.c.g. to the special case called a pure multiplicative generator) and $m=31$ gives, with the starting seed $X_0 = 1$, the sequence $X_1 = 13$, $X_2 = 14$, $X_3 = 27$, $X_4 = 10$, $X_5 = 6$, .... If you have the patience, you can demonstrate for yourself that this sequence has a period of 30 ($=m-1$), i.e., all of the integers from 1 to 30 appear and then repeat. The number 0 can not occur in this generator, of course, because as a multiplicative generator it would then immediately degenerate into producing nothing but zeros.

Since the number 0 can occur in an l.c.g. with $c \neq 0$, the maximum period is $m$ (at the price of having to do an addition for each random number, increasing the period from $m-1$ to $m$ seems a modest gain, indeed). Not all mod $m$ l.c.g.s have this maximum length, however. For example, $a=c=7$ and $m=10$ leads, with $X_0 = 7$, to $X_1 = 6$, $X_2 = 9$, $X_3 = 0$, $X_4 = 7$, $X_5 = 6$, .... The period is only 4. The choices for $a$, $c$, and $m$ are important, then, and a great deal of research has gone into finding "good" values.[2] What "good" means depends, of course, on what properties we want in the resulting sequence.

One obvious property is a big period—so big, in fact, that in any particular application we would never use all of them. That means $m$ should be large; a very popular choice (proposed in 1969) has been $m = 2^{31} - 1 = 2,147,483,647$. When combined with $a = 7^5 = 16,807$ and $c = 0$, we get the pure multiplicative generator that is used in MATLAB in all versions 4.x, prior to Version 5 (5.2 is the latest, as I write, and is the version used in developing all the MATLAB programs given in this book). It has the maximum period, $m - 1 = 2^{31} - 2$; remember, 0 is excluded. This may seem like many numbers but, at a consumption rate of one million numbers per second, we would exhaust the sequence in less than four hours.

Before 1969, some famously bad choices for $a$ and $m$ were widely used; e.g., $a = 65,539$ and $m = 2^{31}$ were used on IBM 360 mainframe computers for a long time. This generator doesn't even have the maximum period. Eventually, that choice was also found to lead to "random" numbers that have very strong correlations among themselves (see below for more about correlation).[3] This problem has caused many of the conclusions arrived at through the use of early l.c.g.s to be called into question; as the authors of an authoritative book wrote, "If all scientific papers whose results are in doubt because of bad [random number generators] were to disappear from library shelves, there would be a gap on each shelf about as big as your fist."[4]

The very latest random number generators, such as MATLAB's Version 5.x, are not based on the l.c.g. algorithm. Rather, they are a combination of shift register/bit manipulation processes that require no multiplication or division operations.[5] Consequently, this new approach is extremely fast. It also leads to incredibly large periods—in Version 5.x of MATLAB, the period is $2^{1492}$, which has led to it being called (somewhat tongue-in-cheek) the "Christopher Columbus" generator. To give you an idea of just how large a number that is, if we consume random numbers at the rate of one million per second, it would take more than $10^{435}$ years (about $10^{425}$ times the age of the universe) before the sequence would repeat. Such generators probably represent the ultimate in random number producers, and they show just how far we have come since 1936, when one mathematician declared, "I cannot conceive that anybody will require multiplications at the rate of 40,000, or even 4,000 per hour."[6]

Three final comments about random number generators: First, the numbers produced are not presented directly to the user, but rather are divided by the modulus ($m$) of the generator; in a maximum period multiplicative generator, each integer from 1 to $m-1$ is divided by m. For MATLAB's Version 4.x, therefore, the smallest and largest possible random numbers are, respectively,

$$1/2,147,483,647 = 0.00000000046566$$

and

$$2,147,483,646/2,147,483,647 = 0.99999999953434.$$

Second, the numbers produced are uniformly distributed over the interval 0 to 1. To illustrate what this means, the MATLAB program **generator.m** (Program 35) creates a histogram of 100,000 numbers produced by Version 5.2's generator. Figure A shows the result for a run using a sequence of 100,000 numbers. The plot does look fairly "flat" (i.e., uniform). All software generators of which I am aware are uniform from 0 to 1 because, from such a generator, it is not difficult to transform the generated numbers into any other desired probability distribution. The theory for how to determine the transformation rule in any particular case can be found in most textbooks on probability theory, but one commonly required transformation is the one that produces numbers uniformly distributed from A to B, where A and B are any two given finite constants $(B>A)$. Most people are willing to accept, at an intuitive level—although it can be derived—that if RAND is a number from a generator uniform from 0 to 1, then $A + (B - A)*RAND$ does the job.

And finally, even more subtle than having a flat or uniform distribution over the unit interval is that a random number generator should (ideally) have zero correlation among the generated numbers themselves. That is, there should be no apparent functional relationship between $x_n$ and $x_{n+j}$, where $j > 0$. This is equivalent to demanding that

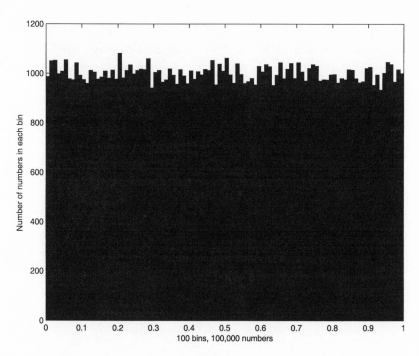

Y-axis label: Number of numbers in each bin

X-axis label: 100 bins, 100,000 numbers

Figure A  Histogram of MATLAB 5.2 Uniform Random Number Generator

if, in a long stream of random numbers, we create the two-dimensional plot of the points $(x_n, x_{n+j})$, we should not see any patterns, such as lines, stripes, lattices, or other regular structures. The MATLAB program **correlation.m** (Program 36) illustrates this for Version 5.2's generator, plotting 1,000 such points (such plots are often called *scatter diagrams*) for any value of $j$ desired. Figure B shows the result for $j = 9$; the plot does not display any obvious patterns that suggest correlation exists or, at the least, the plot indicates that any correlation present is not strong. Now, here is a question for you to ponder (the answer is in note 7, but don't look until you've thought about it for at least a bit):

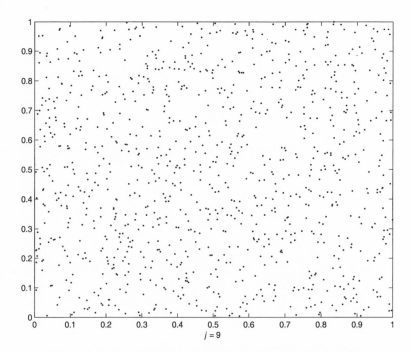

$j = 9$

Figure B  1,000-Point Scatter Diagram for MATLAB 5.2 Uniform Random Number Generator

What do you think the scatter diagram for perfect correlation, i.e., for $j = 0$, would look like?

The *correlation coefficient* between the two random variables U and V is defined to be

$$\rho_{u,v} = \frac{\text{cov } (U,V)}{\sigma_u \, \sigma_v}$$

where $\sigma_u^2 = E\{[U - E(U)]^2\}$ and similarly for $\sigma_v^2$, and the covariance $\text{cov}(U, V) = E\{[U - E(U)][V - E(V)]\}$. If we let $U = X_n$ and $V = X_{n+j}$, then both $X_n$ and $X_{n+j}$ are uniform from 0 to 1 for the MATLAB gen-

erator, with $E(X_n) = (X_{n+j}) = \frac{1}{2}$. Also, $\sigma^2_{x_n} = \sigma^2_{x_{n+j}} =$
$E\{[X_n - \frac{1}{2}]^2\} = E[X_n^2 - X_n + \frac{1}{4}] = E(X_n^2) - E(X_n) + \frac{1}{4}$
$= E(X_n^2) - \frac{1}{2} + \frac{1}{4} = E(X_n^2) = \frac{1}{4}$. Since $E(X_n^2) = \int\limits_0^1 x^2\, dx$
$= \frac{1}{3}$, then $\sigma^2_{X_n} = \sigma^2_{X_{n+j}} = \frac{1}{3} - \frac{1}{4} = \frac{1}{12}$. Thus,

$$\rho_{X_n,\, X_{n+j}} = 12\ \text{cov}(X_n, X_{n+j})$$

where $\text{cov}(X_n, X_{n+j}) = E\{[X_n - \frac{1}{2}]\ [X_{n+j} - \frac{1}{2}]\}$
$= E[X_n X_{n+j} - \frac{1}{2} X_{n+j} - \frac{1}{2} X_n + \frac{1}{4}] = E(X_n X_{n+j} - \frac{1}{4} - \frac{1}{4}$
$+ \frac{1}{4}) = E(X_n X_{n+j}) - \frac{1}{4}$.

Thus,

$$\rho_{X_n,\, X_{n+j}} = 12\ E(X_n X_{n+j}) - 3.$$

Now, for $j \neq 0$, the signature of a "good" generator is
that $X_n$ and $X_{n+j}$ are independent. In that case, it is
easy to show that $E(X_n X_{n+j}) = E(X_n)\ E(X_{n+j}) = \frac{1}{4}$
and so $\rho_{X_n,\, X_{n+j}} = 0$ for $j \neq 0$. For the $j = 0$ case,
$E(X_n X_{n+j}) = E(X_n^2) = \frac{1}{3}$, and so $\rho_{X_n,\, X_n} = 4 - 3 = 1$;
$\rho_{X_n,\, X_n} = 1$ is the mathematical statement that $X_n$ is
perfectly correlated with itself. To test MATLAB's
correlation coefficient performance, I wrote the program cc.m (Program 37), which generates a random
vector X of length 20,009 and then calculates the
value of $12\ E[X(n)X(n + j)] - 3$ for $j = 0$ to 9. Using
this formula does make the implicit assumption that
$X_n$ and $X_{n+j}$ are individually uniform from 0 to 1, but
it does not assume their independence (which is the
point here). The results from one typical run were:

| $j$ | $\rho_{x_n,\ x_{n+j}}$ | $j$ | $\rho_{x_n,\ x_{n+j}}$ |
|---|---|---|---|
| 0 | 1.0230 | 5 | 0.0031 |
| 1 | 0.0103 | 6 | 0.0194 |
| 2 | 0.0236 | 7 | 0.0063 |
| 3 | 0.0184 | 8 | 0.100 |
| 4 | −0.0042 | 9 | 0.0102 |

These deviations from $\rho = 0$ for $j > 0$, and in particular the $\rho > 1$ for $j = 0$ (in theory, $|\rho| \leq 1$), are indications of a generator that deviates (slightly) from the assumed uniformity.

There is one particularly interesting point in the codes for **generator.m** and **correlation.m** to which I want to direct your attention: the statement **rand('state',100\*sum(clock))**. This is a command to use the software clock maintained by MATLAB to compute an initial seed number for the generator. This command is used in all the simulations in this book, and results in an entirely different sequence of random numbers each time a simulation is run (after all, why rerun a simulation if the random numbers are the same, but there is one case where that does makes sense; see the end of the next paragraph).

The variable "clock" is a six-element vector with the format of [year, month, day, hour, minute, second]. For example, if I am writing this on the morning of January 2, 1999, then "right now," my computer's clock vector is [1999, 1, 2, 8, 23, 28.2]. The command "sum" adds the six elements of the vector, and the multiplication by 100 results in a final integer (for the above vector, the integer is 206120) to seed the initial state of the generator. This will obviously be a different integer for each execution of the program. If this forced seeding is not done, then MATLAB automatically uses the same random sequence for every execution (which may, in fact, be helpful during the initial stages of program debugging).

Not all random quantities are uniformly distributed, of course, and perhaps an equally useful distribution for computer simulation work is the so-called normal distribution (or "bell-shaped curve"). The random variable $X$ is said to be normally distributed with mean (average value) $m$ and standard deviation $\sigma$ (or variance $\sigma^2$) if its probability density function (pdf) is

$$f_X(x) = \frac{1}{\sigma\sqrt{2\pi}} \, e^{-\frac{(x-m)^2}{2\sigma^2}}, \; -\infty < x < \infty.$$

The value of $m$ is the location of the peak of the pdf, while $\sigma$ controls the width or spread of the bell. MATLAB has a built-in normal number generator, called **randn**, with zero mean and unit variance. The values of a normally distributed random variable $X$ that has any specified mean $m$ and variance $\sigma^2$ can be generated from MATLAB's generator (call it $Y$) with the simple linear transformation $X = \sigma Y + m$. MATLAB uses a method (unknown to me) to generate normally distributed numbers,[8] but the program **normal.m** (Program 38) plots a histogram of 100,000 values returned by the normal generator; as you can see from Figure C, the histogram is beautifully bell-shaped.

The normal distribution occurs in the theory of measurement errors, a study pioneered by the German mathematician Karl Friedrich Gauss (1777–1855). Because of his enormous contributions to the theory of errors, random quantities that obey the normal law are also called Gaussian random variables. One such quantity, for example, is the *miss distance* of intercontinental ballistic missiles (ICBMs). When launched at a target thousands of miles distant, such missiles will miss their targets by a random amount—specifically, if an arbitrary set of *xy*-axes is drawn with the target at the origin, the $x$ and $y$ miss distances are independent random variables, each with zero mean and equal variances. The radial miss distance, which is what really counts as far as warhead blast effects are concerned, is then the random variable $R = \sqrt{X^2 + Y^2}$, where $X$ and $Y$ are the $x$ and $y$ miss distances. $R$ is Rayleigh distributed, a probability density named after its nineteenth-century dis-

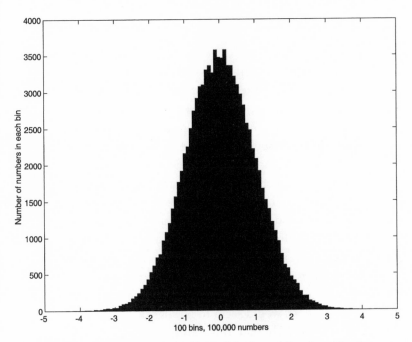

Figure C  Histogram of MATLAB 5.2 Normal Random Number Generator

coverer, Lord Rayleigh (the Englishman John William Strutt (1842–1919)).[9]

Let's now assume that we have a source for all the random numbers we could ever hope to need. What can we do with them? To end this chapter, I'll show you three examples of how to simulate physical processes, using random numbers, that may be too hard to do theoretically. The three examples I'll use are actually not very hard to solve theoretically, so we'll have something to check our simulation results against.

For my first example, I'll return to Marilyn vos Savant's column in *Parade* magazine (see the introduction), and, in particular, to the one from the December 27, 1998, issue.

There, she printed the following letter from one of her readers:

> At a monthly "casino night," there is a game called Chuck-a-Luck. Three dice are rolled in a wire cage. You place a bet on any number from 1 to 6. If any one of the three dice comes up with your number, you win the amount of your bet. (You also get your original stake back.) If more than one die come up with your number, you win the amount of your bet for each match. For example, if you had a $1 bet on number 5, and each of the three dice came up with 5, you would win $3. It appears that the odds of winning are 1 in 6 for each of the three dice, for a total of 3 out of 6—or 50%. Adding the possibility of having more than one die come up with your number, the odds would seem to be slightly in the gambler's favor. What are the odds of winning at this game? I can't believe that a casino game would favor the gambler.

Marilyn's answer wasn't wrong but, in my opinion, it also wasn't particularly enlightening. I do wonder, too, why she said nothing about the incorrect assertion that if one die has a 1 in 6 chance of winning, then 3 dice have a 50 percent chance of winning. Does that mean 9 dice have a 150 percent chance of winning? Of course not. To see how to properly calculate the correct assertion, take another look at the discussion, in Part II of the introduction, of Marilyn's June 14, 1998, column. The probability of winning with a roll of three dice is $1 - \left(\frac{5}{6}\right)^3$ or 42.13 percent, not 50 percent, and the probability of winning with a roll of nine dice is $1 - \left(\frac{5}{6}\right)^9$ or 80.62 percent, not the absurd 150 percent.)

After a few specific examples of how the casino does

win, Marilyn concludes by saying that "with this game, you can expect to lose about 8 cents with every \$1 you bet over the long run." She offers no proof of this (she is right, but I do wonder if she really knows how to calculate the result, or if she just found it in some uncited book), so here is how to do it.

You lose \$1 (your bet) if none of the three dice shows your number. This happens with probability $\left(\frac{5}{6}\right)^3 = 125/216$. Alternatively, you win $-\$1$ with this probability.

You win \$1 if exactly one die shows your number. This happens with probability $3(1/6)(5/6)^2 = 75/216$.

You win \$2 if exactly two dice show your number. This happens with probability $3(1/6)^2(5/6) = 15/216$.

You win \$3 if all the dice show your number. This happens with probability $(1/6)^3 = 1/216$. So, your average winnings (in dollars) are $(-1)(125/216) + (1)(75/216) + (2)(15/216) + (3)(1/216) = -(17/216) = -0.0787$. That is, on average you lose just under 8 cents for each dollar you bet.

That's the theoretical answer; what would a simulation predict? The MATLAB program **casino.m** (Program 39) plays 10,000 virtual games of Chuck-a-Luck by using the random number generator to select four integers, each from 1 to 6. One number is your bet, and the other three are the faces that show on the dice. This is done with the "floor" command, which rounds its argument toward minus infinity, e.g., floor(3.2) = 3 and floor($-3.2$) = $-4$. So, since **6\*rand** is between 0 and 6, then **floor(6\*rand)** is between 0 and 5, and thus **floor(6\*rand)** + 1 is between 1 and 6. The program determines how many matches (if any) there are between the bet and the dice, and thus how much you've won (or lost) on that play, and uses that result to update the variable $W$, your total winnings. When I ran **casino.m** three times, the results were $-0.0945$, $-0.0864$, and $-0.0751$, which is in fairly good agreement, I think, with the theory.

For my second example, consider the "Thief of Baghdad" problem. It has a somewhat tricky theoretical solution, but its simulation is just as easy as the one for Chuck-a-Luck. When thieves were captured in old Baghdad (so goes this tale), they were thrown into a diabolical prison. A thief would find himself in a pitch-black, round room with a perfectly smooth circular wall. Spaced equally around the wall were three identical doors. Once a thief passed through a door, it automatically locked behind him. One door led, after a short walk in the dark, immediately to freedom. Another door led, after a short walk in the dark, to a twisty tunnel that took $S$ hours to crawl through, only to drop the thief through a hole in the ceiling back into the prison room. And the final door did the same, after a crawl of $L$ hours.

Since a thief was always completely disoriented after each tunnel crawl and ceiling drop, he would pick a door for his next escape attempt at random; i.e., the probability of selecting each door was always 1/3. It is conceivable, of course, for a thief to have been stuck in this maddening prison for an arbitrarily long time. So, a natural question is: If many thieves were so imprisoned, how long would it take, on average, to escape?

Let $E$ be the answer. If the thief took the $S$ tunnel, then once back in the room he would, *at that time*, again expect it would take additional time $E$ to escape. The same argument holds for the case of taking the $L$ tunnel. If he took the door to freedom (which takes essentially zero time) he then expected zero time to escape, as he was already free.) So,

$$E = (1/3)(S + E) + (1/3)(L + E) + (1/3)(0 + 0)$$

or, with very little algebra, we arrive at the amazingly simple result of $E = S + L$.

This clever little trick is actually more than a trick; it is a method, which is a trick that you can use more than once. For example, see if you can apply it to solve the following puzzle: An integer is selected at random from the set 1, 2, ... , $N$; i.e., each has probability $1/N$ of being selected. Call the selected integer $X_1$. Then, repeat the selection process for the set 1, 2, ... , $X_1$, with each integer now having probability $1/X_1$ of being selected. Continue to do this until the selected integer is $X_M = 1$. The value of $M$, the number of selections until the selected integer is 1, is clearly a random quantity. What is the average value of $M$?

The solution is given in the box at the end of the book, but don't peek until you've tried hard for a while. To help get you started, observe that the answer will be some function of $N$ (this is not as trivial an observation as it might seem). So let's write the answer, $E(M)$, as

$$E(M) = f(N).$$

Whatever the function $f$ is, it is trivially obvious that $f(1) = 1$. What is $f(2)$? If $N = 2$, then we can write the following:

$M = 1$ with probability $\frac{1}{2}$; $X_1 = 1$

$M = 2$ with probability $\left(\frac{1}{2}\right)^2$; $X_1 = 2$, $X_2 = 1$.

$M = 3$ with probability $\left(\frac{1}{2}\right)^3$; $X_1 = X_2 = 2$, $X_3 = 1$.

$M = j$ with probability $\left(\frac{1}{2}\right)^j$; $X_1 = X_2 = ... = X_{j-1}$
$= 2$, $X_j = 1$.

So, given that $N = 2$, we can write

$$E(M) = 1\left(\frac{1}{2}\right) + 2\left(\frac{1}{2}\right)^2 + 3\left(\frac{1}{2}\right)^3 + \ldots + j\left(\frac{1}{2}\right)^j + \ldots = 2.$$

(If you don't see how to do this sum, take a look at the solution for problem 15.) Thus, $f(2) = 2$.

Repeating this approach for $N \geq 3$ might be a possibility, but rather than doing that, see if you can now figure out how to use the "Thief of Baghdad" trick to find a formula for $f(N)$. It should, of course, give $f(1) = 1$ and $f(2) = 2$. And no, the answer is *not* $f(N) = N$.

The MATLAB program **thief.m** (Program 40) simulates the escapes of 10,000 virtual thieves, for the case of $S = 1$ and $L = 3$ (the variable names are, respectively, "short" and "long"). The theoretical answer is, of course, 4; when I ran **thief.m** three times, the results were 4.0126, 3.9735, and 3.9861. Again, the agreement between theory and simulation is not bad.

As my final example, consider the following problem that has become famous among mathematicians in recent years, the so-called Monty Hall problem, named after the host of the popular television game show "Let's Make a Deal." If you were a contestant on that program, Monty would show you three doors, of which you would select one. Behind one of the doors was a valuable prize, and behind the other two were goofy joke prizes. After you had made your selection, Monty would open the door of one of the joke prizes (he knew, of course, what was behind each door), and offer you the opportunity to switch; i.e., to select the other unopened door. And that's the puzzle:

Should you switch doors or not? It intuitively seems to most people (including me) that it should make no difference whether you switch or not—but it does. As one recent paper that discusses this problem puts it,

> You have a $\frac{1}{3}$ chance of initially picking the [door] with the prize—and therefore a $\frac{2}{3}$ chance of initially *not* picking the [door] with a prize. If you have picked the prize-containing [door] and keep it, you win—with a probability of $\frac{1}{3}$. If you have not initially picked the prize-containing [door] and switch, you win—with a probability of $\frac{2}{3}$. Thus the odds of winning are twice as great if you switch.[10]

Despite that argument, many people (including me) still feel a bit unsure of such a dramatic conclusion. This is a problem, then, that demands a computer simulation, one performed by the MATLAB program **monty.m** (Program 41). That program plays 10,000 games of the Monty Hall problem, and keeps track of how many games you would win by following each of the two strategies, *switch* or *don't switch*. The results of three runs were:

|  | Games Won (out of 10,000) | |
| --- | --- | --- |
|  | Switch Strategy | Don't Switch Strategy |
| Run 1 | 6,621 | 3,379 |
| Run 2 | 6,713 | 3,287 |
| Run 3 | 6,700 | 3,300 |

These results confirm the validity of the switch strategy, as well as the doubling of the probability of winning with that strategy. It's strange, but true.

# Notes and References

1. Such tables were actually produced. A famous one is the 600-plus-page-long *A Million Random Digits with 100,000 Normal Deviates*, published in 1955 by the RAND Corporation in Santa Monica, California. (The name is not a joke—RAND is an acronym for "Research and Development," not "random." Project RAND originated in 1946 as an advisory agency within the Douglas Aircraft Company for the soon-to-be-created Air Force.) All those digits could be purchased from RAND on a deck of punched cards, too (an obsolete technology, like punched paper tape—a spilled coffee cup next to punched paper was a disaster—that my young students have a hard time believing was once the industry standard) for automatic machine use. Modern Monte Carlo practitioners can only shake their heads with a mixture of amusement and admiration at the Rube Goldberg way in which those numbers were produced. As the introduction tells us: "Briefly, a random frequency pulse source, providing on the average about 100,000 pulses per second, was gated about once per second by a constant frequency pulse. Pulse standardization circuits passed the pulses through a 5-place [i.e., 5-bit, in modern terminology] binary counter. In principle, the machine was a 32-place roulette wheel which made, on the average, about 3000 revolutions per trial and produced one number per second. A binary-to-decimal converter was used which converted 20 of the 32 numbers (the other twelve were discarded) and retained only the final digit of two-digit numbers; this final digit was fed into an IBM punch to produce finally a punched card table of random digits." Modern interest in such a book has understandably narrowed to just historians. When I checked a copy of the RAND book out of the University of New Hampshire's Dimond Library, for example, while writing this chapter, I couldn't help but notice that the last person before me had taken the book out thirty-three years ago (in 1965).

2. See the first chapter of Donald Knuth's *Seminumerical Algorithms*

(Reading, Massachusetts: Addison-Wesley, 1969). That chapter is devoted to nothing but the theory of random number generation, and at 160 pages it is (despite being thirty years old) still a wonderful source of a vast amount of useful information.

3. For details see the famous paper by George Marsaglia, "Random Numbers Fall Mainly in the Planes," *Proceedings of the National Academy of Sciences* 61 (September 15, 1968):25–28. A more recent paper is F. James, "A Review of Pseudorandom Number Generators," *Computer Physics Communications* 60 (1990):329–344.

4. William H. Press et al., *Numerical Recipes in FORTRAN: The Art of Scientific Computing*, 2d ed. (New York: Cambridge University Press, 1992), p. 267.

5. For a discussion of how this works, see George Marsaglia et al., "A Random Number Generator for PC's," *Computer Physics Communications* 60 (1990):345–349.

6. Quoted in Knuth (note 2), p. 161.

7. A straight line with slope $+1$ (a plot with no scatter).

8. When I asked the distributors of MATLAB (The MathWorks in Natick, Massachusetts) about the normal generator, they either couldn't or wouldn't tell me (perhaps the details are proprietary). In any case, there are a number of theoretically perfect ways to do it; here's one that is based on the availability of a uniform generator. If $U$ and $\theta$ are independent random variables, uniform from 0 to 1 and 0 to $2\pi$, respectively, then the pair $X_1 = \sqrt{-2\ln(U)}\cos(\theta)$ and $X_2 = \sqrt{-2\ln(U)}\sin(\theta)$ are independent random variables that are each Gaussian with zero mean and unit variance. You can find a very terse sketch of a proof of this assertion (which will mean nothing to you unless you are familiar with Jacobian determinants—see also the end of Problem 14—of which the proof says not a word) in a paper by G. E. P. Box and Mervin E. Müller, "A Note on the Generation of Random Normal Deviates," *The Annals of Mathematical Statistics* 29 (1958): 610–611. An extended discussion of these equations (along with mention of Jacobian determinants), with more recent enhancements, can be found in the book by John Dagpunar, *Principles of Random Variate Genera-*

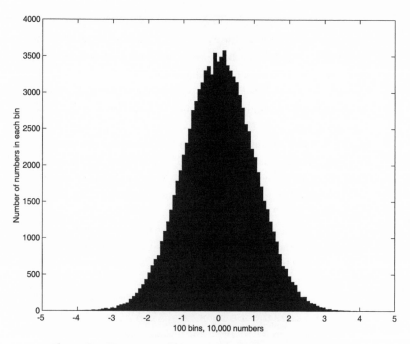

*Figure* D   Histogram of Normal Random Number Generator

*tion* (New York: Oxford University Press, 1988), pp. 93–95. The MATLAB program **onewaytodoit.m** (Program 42) generates 10,000 random numbers using the original Box and Müller equations and plots (see Figure D) a histogram. It is, indeed, bell-shaped.

9. The random variable $R$ is Rayleigh if its probability density function is

$$f_R\left(r\right) = \frac{r}{\sigma^2}\,e^{-r^2/2\sigma^2},\, r \ge 0,$$

and zero otherwise. It was discovered in 1887, in connection with Lord Rayleigh's theoretical studies in the theory of sound.

10. Mark P. Silverman, Wayne Strange, Chris R. Silverman, and Trevor C. Lipscombe, "On the Run: Unexpected Outcomes of Random Events," *The Physics Teacher* 37 (April 1999): 218–225. More

discussion of the Monty Hall problem can be found in the interesting book by H. W. Lewis, *Why Flip a Coin? The Art and Science of Good Decisions* (New York: John Wiley & Sons, 1997), pp. 191–177. As I write this, an amusing animated simulation of the Monty Hall problem is on the Web at http://www.intergalact.com/threedoor/threedoor.html.

"Some Things Just Have to Be Done by Hand!"

This story originally appeared, in slightly different form, in Analog Science Fiction Magazine's *Analog Yearbook 1984*, edited by Stanley Schmidt.

The Most Important Entity rubbed His temples in fatigue. There was just so damned much crap to put up with nowadays. The personnel paperwork was nearly overwhelming, even for a being with omnipotent powers. And a work force faced with zero turnover had a first-class morale problem. The younger ones knew there was no hope for advancement by the once-usual routes of death, retirement, or resignation. None of those events ever happened—here.

The telephone rang, and He answered in weary relief at the distraction. "Yes?"

"Sorry to bother you, Sir, but the main computers have a backlog in the RANDOM QUEUE for ten to the 183d power decisions. Can you please service those requests right now?"

"Damn, are those bloody scientists on Earth doing their quantum experiments again!? You'd think they'd understand the Uncertainty Principle after all these years. Well, what is it now, an electron beam through a diffraction grating, or is somebody trying to locate an atom with zero error?"

"Both, and more, Sir. Those guys are really getting busy down there. Why, just as we've been talking here, the RQ has picked up ten to the 179th power more requests!"

The main computers couldn't be allowed to overflow. Once, two or three thousand years ago (in Earth time), they had been unattended for several days (in His time), and the RQ had clogged up tight with ignored decision requests for determining the outcomes of random events. The resulting massive computer system crash had caused entire centuries (in Earth time) of strange, abnormal violations in His Laws of Natural Phenomena. It had been the time of magic on earth, and the new wizards, sorcerers, and magicians had used it to their advantage in proclaiming themselves all-powerful. It couldn't be allowed to happen again!

"All right, all right, hold your feathers smooth. Hang on for a moment." He put His caller on hold, and pulled open the desk drawer next to His perfect left foot. Inside was a pure diamond crystal box, containing two ruby cubes of ultimate clarity. The dots on the cube faces were precise circles of gold. Taking the cubes in His mighty hand, He established a mind-link with the input-output data lines to the main computers. Faster than imaginable (or even possible by ordinary laws, but for Him very little was impossible) the cubes tumbled in His quivering hand. The whole thing was over in just a few wingbeats.

"Okay, the main computers cleaned up?"

"Yes Sir, the RANDOM QUEUE is empty!"

"Excellent—now please don't call again for at least another day. Meanwhile, you and your colleagues might busy yourselves with finding a way to speed up the automatic software random number generator. I find this business of hand-generation to be increasingly inconvenient! Good-bye."

As He hung up, He thought of what Albert Einstein, one of the better Earth scientists, had once said: "God doesn't play dice with the Cosmos."

"Hummph," He grunted in disgust to Himself, "just what the Hell did *he* know about it?"

## Solution to the Puzzle in the Random Numbers Chapter

Just before the first selection, when we expect there to be $f(N)$ selections, we will pick either $N$, or $N - 1$, or $N - 2, \ldots$ , or 2, or 1, with each possibility having probability $1/N$. If the selection is $N$, we are right back with the original situation, but we have used up a try. Thus, with probability $1/N$, the expected number of selections is $[1 + f(N)]$. If the selection is $N - 1$, then we have used up a selection and we "expect" there to be $f(N - 1)$ more selections; i.e., with probability $1/N$, the expected number of selections is $[1 + f(N - 1)]$. We can continue to reason this way—which is, of course, the "Thief of Baghdad" trick—all the way down to selecting 2. Finally, if the initial selection is 1, then we have used up a selection and we are done (which happens with probability $1/N$). So, for $N \geq 2$, we can write

$$f(N) = \frac{1}{N}[1 + f(N)] + \frac{1}{N}[1 + f(N - 1)] + \ldots$$

$$+ \frac{1}{N} [1 + f(2)] + \frac{1}{N} [1].$$

With a little algebra, this becomes

$$f(N) = \frac{1}{N-1} \left[ N + \sum_{j=2}^{N-1} f(j) \right], N \geq 2.$$

Notice that this does indeed reduce to $f(2) = 2$.

From this result, we can now calculate f(N) for any N, e.g.,

$$f(3) = \frac{3 + f(2)}{3 - 1} = \frac{3 + 2}{2} = 2.5;$$

$$f(4) = \frac{4 + f(3) + f(2)}{4 - 1} = \frac{4 + 2.5 + 2}{3} = 2.8333;$$

$$f(5) = \frac{5 + f(4) + f(3) + f(2)}{5 - 1} = \frac{5 + 2.8333 + 2.5 + 2}{4}$$

$$= 3.0833.$$

It is easy to verify by direct computation that a more elegant expression for $f(N)$ is

$$f(N) = 1 + \sum_{j=1}^{N-1} \frac{1}{j}, N = 1, 2, 3, \ldots$$

The sum is, of course, just the first $N - 1$ terms of the harmonic series, which has been known since the middle of the fourteenth century to diverge. But the divergence is very slow; for $f(N)$ to exceed 16, N must be larger than 1.6 million.

# MATLAB Programs

# 1. 'idiots1.m

```
%idiots1.m/created by PJNahin for "Duelling Idiots"(7/3/98)
%This m-file simulates the original dueling idiots. The elements
%of the row vector duration are the number of duels of length k,
%i.e., duration(k)=# of duels that require k trigger-pulls to
%complete, where k=1,2,3,....
%
%
Duels=10000;                       %total number of duels;
duration=zeros(1,60);              %assume no duel exceeds 60 trigger-pulls;
a=0;                               %number of times A has won, so far;
rand('state',100*sum(clock))       %new seed for generator
nd=0;                              %number of duels completed, so;
tp=0;                              %number of trigger pulls, so far, in present duel
while nd<Duels
    ra=rand;                       %get two random numbers
    rb=rand;
    tp=tp+1;                       %A pulls the trigger
    if (ra<=1/6)                   %A wins
        duration(tp)=duration(tp)+1;   %duel is over, up-date the
                                       %duration vector;
        tp=0;                      %initialize number of trigger-pulls for
                                   %the next duel;
        a=a+1;
        nd=nd+1;
    else
        tp=tp+1;                   %B pulls the trigger
```

# 1. idiots1.m (cont.)

```matlab
    if (rb<=1/6)        %B wins
        duration(tp)=duration(tp)+1;    %duel is over, up-date the
                                        %duration vector;
        tp=0;                           %initialize the number of
                                        %trigger-pulls for the
                                        %next duel;

    nd=nd+1;

    end

end                 %start next duel;
a=a/Duels;
disp(['The probability A wins is ',num2str(a)])
average=0;
for k=1:length(duration)
    average=average+k*duration(k);
end
average=average/Duels;
disp(['The average number of trigger-pulls/duel is ',num2str(average)])
bar(duration)       %plot a bar graph of the duration vector
title('Figure 2.1 - Relative Frequency of the Number of Trigger-Pulls per Duel')
xlabel('duration of duels, in units of trigger-pulls')
ylabel('number of duels')
figure(1)
```

## 2. pisim.m

```
%pisim.m/created by PJNahin for "Duelling Idiots"(10/22/98)
%This m-file estimates pi by randomly tossing darts at the
%unit square and counting how many land inside a quarter-
%circle with unit radius contained in the square.
%
%
rand('state',100*sum(clock))    %set new seed for generator;
darts=0;                        %initialize number of darts
                                %inside quarter-circle region;

for i=1:10000;
   x=rand;                      %toss a
   y=rand;                      %dart;
   r=x*x+y*y;                   %compute distance squared from origin to dart;
   if r<1                       %is dart inside quarter-circle region?
      darts=darts+1;            %yes
   end
end
pi_estimate=4*darts/10000
```

3. gas.m

```
%gas.m/created by PJNahin for "Duelling Idiots"(5/22/99)
%This m-file simulates the diffusion of gas molecules in a sealed
%container by using the Ehrenfest ball exchange rules. The simulation
%starts with n molecules (i.e., balls) of one type (i.e., black) on
%one side of the container, and n more molecules of another type (i.e.,
%white balls) on the other side. The two urns play the roles of the
%two sides of the container. To simulate the ball (molecule)
%movements, the program selects two random numbers from 0 to 1, which
%are then compared to the current probabilities of selecting a black
%ball from urn I and a white ball from urn II. If BOTH random numbers
%are greater than these two probabilities then a white ball has been
%selected from urn I and a black ball has been selected from urn II,
%and so the number black balls in urn I is increased by one while the
%number of white balls in urn II is increased by one. If BOTH random
%numbers are less than or equal to these two probabilities then a
%black ball has been selected from urn I and a white ball has been
%selected from urn II and so the number of black balls in urn I is
%decreased by one while the number of white balls in urn II is decreased
%by one. If one of the random numbers is greater than its corresponding
%probability while the other random number is less than its
%corresponding probability, then no action is taken because then a
%white (black) ball moves from urn I to urn II at the same time a white
%(black) ball moves in the opposite direction. That is, there is no
%net change. Then, the ball selection probabilities are recalculated and
%another ball exchange is simulated.
%
%
```

## 3. gas.m (cont.)

```matlab
rand('state',100*sum(clock))        %new seed for the generator;
n=50;                               %number of balls in each urn;
nb1=n;                              %number of black balls INITIALLY in urn I;
nw2=n;                             %number of white balls INITIALLY in urn II;
pb1=nb1/n;                          %probability of selecting a black ball from urn I;
pw2=nw2/n;                          %probability of selecting a white ball from urn II;
for trials =1:600;
    system(trials)=pb1;
    ball1=rand;
    ball2=rand;
    if(ball1>pb1&ball2>pw2)         %white ball selected from urn I
        nb1=nb1+1;                  %and black ball selected from
        nw2=nw2+1;                  %urn II;
    elseif(ball1<=pb1&ball2<=pw2)   %black ball selected from urn I
        nb1=nb1-1;                  %and white ball selected from
        nw2=nw2-1;                  %urn II;
    end
    pb1=nb1/n;
    pw2=nw2/n;
end
plot(system)
axis([1 trials 0 1])
grid
xlabel('time, in arbitrary units')
ylabel('fraction of balls in urn I that are black')
title('Fig.8.1 - Simulation of the Ehrenfest Ball Exchange Process')
figure(1)
```

```matlab
%odd.m/created by PJNahin for "Duelling Idiots"(7/1/98)
%This m-file simulates 1,000 games of odd-person-out, for N people as
%defined by the user, with each person flipping a fair coin. The elements
%of the row vector duration are the number of games of length i, i.e.,
%duration(i)=# of games that require i flips to complete, where i=1,2,3...
%
%
duration=zeros(1,50);              %initialize to zero the number of games
                                   %that have durations from 1 to 50 flips
N=input('Enter number of people playing the game:');
rand('state',100*sum(clock))       %set new seed for the random number
                                   %generator
for game=1:1000                    %simulate 1,000 games
    gameover=0;                    %gameover=1 is the flag that the
                                   %current game is done
    flip=1;                        %initialize duration of current game
    while gameover==0              %keep playing current game
        s=0;                       %initialize number of heads before
                                   %each flipping of N coins
        for n=1:N                  %flip a fair coin for each of the N
            coin=rand;             %people, and increment s by 1 for
            if coin<1/2            %each head
                s=s+1;
            end
        end
        if  s==1 | s==N-1          %is there an odd-person-out?
```

## 4. odd.m (cont.)

```matlab
            gameover=1;          %yes, set game-over flag
    else
            flip=flip+1;         %no, so flip again
    end                          %check while statement for end of game
    duration(flip)=duration(flip)+1; %game is over, up-date duration vector
end                              %play new game
average=0;
for i=1:length(duration)
    average=average+i*duration(i);
end
average=average/1000;
disp(['The average number of flips/game for ',num2str(N),' players is ',num2str(average)])
bar(duration)                    %plot a bar graph of duration vector
title('Fig.9.1-Relative Frequency of the Duration of Odd-Person-Out')
xlabel('duration of games, in units of flips')
ylabel('number of games')
```

```
%brownian.m/created by PJNahin for "Duelling Idiots"(3/8/99)
%This m-file simulates Brownian motion (a random walk by a particle
%suspended in a fluid and being hit by molecules). At successive
%increments of time the particle performs independent motions of random
%length, uniform from -1 to 1 (in arbitrary units) in both the x and the y
%directions. The walk duration (the total number of steps, where a 'step'
%is a delta-x and a delta-y *pair*) is defined by the user. A walk always
%starts at the origin.
%
%
n=input ('Walk duration (in steps)? ');
rand('seed',100*sum(clock))
clear xpos
clear ypos
xpos(1)=0;
ypos(1)=0;
for i=2:n
    deltax=-1+2*rand;
    deltay=-1+2*rand;
    xpos(i)=xpos(i-1)+deltax;
    ypos(i)=ypos(i-1)+deltay;
end
plot(xpos,ypos)
xlabel('x direction')
ylabel('y direction')
title('Fig.19.2 - Two-Dimensional Brownian Motion')
figure(1)
```

6. cpm.m

```matlab
%cpm.m/created by PJNahin for "Duelling Idiots"(10/17/98)
%This m-file executes the Critical Path Method algorithm
%for the project shown in Figure 21.1.
%
%
ntasks=8;                    %number of tasks, where task #1 is the pseudo-task
                             %BEGIN and the last task is the pseudo-task END;
                             %it is assummed that all tasks are numbered in
                             %sequence, from BEGIN as task #1 to END;
                             %enter individual task data, in the following
                             %format: for task i, data(i,:)=
                             %[<# of immediate predecessor tasks>,
                             %<task #'s of immediate predecessors>,
                             %<time to do task>];
                             %IMPORTANT POINT 1: all data statements must be
                             %of the same length, so pad-out all statements
                             %that have less than the maximum length with
                             %zeros.
                             %IMPORTANT POINT 2: the data statements MUST be
                             %entered in sequence from BEGIN to END, i.e.,
                             %all of the predecessor tasks of each task
                             %must appear in earlier data statements;

data(1,:)=[0,0,0,0];
data(2,:)=[1,1,2,0];
data(3,:)=[1,1,1,0];
data(4,:)=[2,2,3,1];
data(5,:)=[2,2,3,2];
data(6,:)=[1,4,2,0];
data(7,:)=[2,5,6,1];
data(8,:)=[1,7,0,0];

%
time=zeros(50,5);            %initialize matrix of important times;
p=zeros(50,50);              %initialize immediate predecessor matrix;
for i=1:ntasks               %for task i, get number of immediate predecessors
```

```
npred=data(i,1);              %where data(i,2) to data(i,1+npred) are the task
                              %numbers of the immediate predecessors of task i;

k=1+npred;
  for j=2:k                   %tag the task pointed at by data(i,j) as
    p(i,data(i,j))=1;         %an immediate predecessor of task i, i.e.,
                              %p(i,q)=1 if task q is an immediate
                              %predecessor of task i;
  end
time(i,5)=data(i,k+1);        %enter time to do task i into
                              %matrix of important times
end
%Start forward pass to determine the earliest start and finish times
%for each task, stored as time(i,1) and time(i,2), respectively.
for i=1:ntasks
  earlyctime=0;               %initialize earliest possible completion
                              %time for immediate predecessors of task i
                              %(we are looking for the immediate predecessor
                              %with the MAXIMUM completion time, as task i
                              %cannot be started until ALL of its predecessors
                              %are done);

  for j=1:ntasks
    if p(i,j)==1              %if task j is an immediate
                              %predecessor of task i then ......
      if earlyctime<time(j,2) %if we have found a new
                              %maximum completion time then ......
```

```matlab
            earlyctime=time(j,2);          %up-date maximum
                                           %completion time of a
                                           %predecessor task to task i;

        else
        end
    else
    end
                                           %have we checked for all
                                           %predecessor tasks?
    time(i,1)=earlyctime;                  %yes, so record earliest time task i can
    time(i,2)=earlyctime+time(i,5);        %start and record earliest time task i
                                           %can be finished;
end                                        %repeat for next task;
%
%
%Start backward pass to determine the latest start and finish times
%for each task, stored as time(i,3) and time(i,4), respectively.
projectdone=time(ntasks,1);                %get the time required to do the project
for i=ntasks:-1:1
    lateftime=projectdone;                 %initialize latest possible finish
                                           %time for task i; to do this, look
                                           %at all of the successor tasks for
                                           %task i and find the one that starts
                                           %soonest, i.e., find the successor
                                           %task that has the MINIMUM start
                                           %time (if task i starts later than
                                           %that time then that successor task
                                           %will be delayed);
```

6. cpm.m (cont.)

```matlab
  for j=1:ntasks
    if p(j,i)==1;                 %if task i is an immediate predecessor
                                  %task j (i.e., if task j is an
                                  %immediate successor of task i) then ...
      if lateftime>time(j,3)      %if we have found a new minimum
                                  %start time for task j then ...
        lateftime=time(j,3);      %up-date late finish time
      else                        %for task i;
      end
    else
    end
  end                             %have we checked for all successor tasks?
  time(i,4)=lateftime;            %yes, so record latest time task i can end;
  time(i,3)=lateftime-time(i,5);  %and latest time that task i can start;
end
%Display results
disp('Total time required for project = '), disp(projectdone)
disp('task    early start time    slack')
realtasks=ntasks-1;
for i=2:realtasks
  slack=time(i,3)-time(i,1);
  ans=[i    time(i,1)         slack];
  fprintf('  %g        %g              %g\n',ans(1),ans(2),ans(3))
end
```

7. 'idiots2.m

```matlab
%idiots2.m/created by PJNahin for "Duelling Idiots"(7/6/98)
%This m-file evaluates the theoretical probability that idiot A
%wins the duel with idiot B, using the revised duelling rules
%described in the problem assignment. The calculations are done
%in symbolic form to avoid round-off errors (see the definition
%of the variable r below). The final result is printed to 26 digits
%and is printed on the screen after each additional term is added
%to allow watching the answer converge.
%
%

S=0;                    %initialize finite geometric series sum that
                        %appears in each term;
T=1;
r=sym('5/6');           %the first term in the brackets for P(A);
e=3;

incr=7;                 %starting value of exponent in the factor that
                        %multiplies the finite geometric series;
                        %initial amount to be added to exponent to get
                        %the next exponent;
startp=0;               %initial value of power in the first term of
                        %the present finite geometric series;
stopp=2;                %initial value of power in the last term of
                        %the present finite geometric series;
for k=1:30              %the '30' is somewhat arbitrarily picked to
                        %be sufficiently large to see convergence;
    for i=startp:stopp  %form finite geometric sum from old sum by
                        %adding-on two new terms (the FIRST sum has
                        %THREE terms);
```

# 7. idiots2.m (cont.)

```matlab
    S=S+r^i;
  end
M=r^e;                    %factor to multiply geometric series;
T=T+M*S;                  %multiply, and add to old total;
startp=stopp+1;           %up-date starting power for new terms to add
                          %to old sum;
stopp=startp+1;           %up-date final power for new terms to add
                          %to old sum;
e=e+incr;                 %up-date exponent of factor to multiply
                          %with new geometric series;
incr=incr+4;              %up-date exponent increment;
vpa(T/6,26)               %print variable precision arithmetic value (26
                          %digits) of present value for P(A);
                          %terminate the program via the keyboard when
                          %convergence is observed;
end
```

## 8. bulb.m

```
%bulb.m/created by PJNahin for "Duelling Idiots"(5/9/98)
%This m-file calculates and plots the probability of a bulb,
%wired in series/parallel with sheets of switches, glowing.
%
%
n=input('Number of switches in a row? ')
for p=1:200
    r=p/200;
    P1(p)=(1-(1-r^n)^n)^n;
    P2(p)=1-((1-r^n)^(n*n));
end
r=.005:.005:1;
plot(r,P1,'-',r,P2,'.')
grid
title('Fig.3.4-Solid Line for Series Sheets, Dots for Parallel Sheets')
xlabel('probability, p, an individual switch is closed')
ylabel('probability bulb glows, P(10,p)')
figure(1)
```

# 9. markov.m

```
%markov.m/created by PJNahin for "Duelling Idiots"(3/20/99)
%
p0=[1 0 0 0];                           %make initial state vector p0 a row vector;
r1=[.97 .03 0 0];
r2=[0 .98 .02 0];
r3=[0 0 .99 .01];
r4=[0 0 0 1];
A=[r1;r2;r3;r4];
B=A;
for n=1:800
    p=p0*B;                             %form new state vector;
    C=A*B;
    B=C;
    prob3(n)=p(4);                      %save new state vector elements in
    prob2(n)=p(3);                      %prob* (where * = 4, 3, 2, 1);
    prob1(n)=p(2);
    prob0(n)=p(1);
end
n=1:800;
plot(n,prob3,n,prob2,'.',n,prob1,'+',n,prob0,'*')
legend('state 3','state 2','state 1','state 0')
xlabel('time n (in microseconds)')
ylabel('probability')
title('Fig.3.5-Probability the Path is in State k at Time n')
grid
figure(1)
```

## 10. underdog1.m

```
%underdog1.m/created by PJNahin for "Duelling Idiots"(2/17/98)
%
%
for p1=50:100
    p=p1/100;
    common=(1-p)^4;
    weaker=common*(1+4*p+10*p*p+20*p*p*p);
    P(p1-49)=weaker;
end
p=.5:.01:1;
plot(p,P)
grid
xlabel('probability, p, of stronger team winning an individual game')
ylabel('probability weaker team wins the World Series')
title('Fig.4.1-The Stronger Team Does NOT Always Win!')
figure(1)
```

## 11. underdog2.m

```
%underdog2.m/created by PJNahin for "Duelling Idiots"(2/17/98)
%
%
%
duration=zeros(1,100);
for i=1:100
    p=i/100;
    common1=(1-p)^4;
    common2=p^4;
    duration(i)=4*(common1+common2)+20*(common1*p+common2*(1-p));
    duration(i)=duration(i)+60*(common1*p^2+common2*(1-p)^2);
    duration(i)=duration(i)+140*(common1*p^3+common2*(1-p)^3);
end
p=.01:.01:1;
plot(p,duration)
grid
xlabel('Probability, p, of a team winning an individual game')
ylabel('Duration, in games')
title('Fig.4.2-Expected Duration of the World Series')
figure(1)
```

## 12. esim.m

```
%esim.m/created by PJNahin for "Duelling Idiots"(12/20/98)
%This m-file estimates e by performing the simulation described
%in the statement of Probblem 5.
%
rand('state',100*sum(clock));          %new seed for generator;
N=input('What is N? ')
bin=zeros(1,N);
for j=1:N
    number=floor(N*rand)+1;            %floor rounds towards minus infinity;
    bin(number)=bin(number)+1;
end
empty=0;
for j=1:N
    if bin(j)==0                        %empty = # of empty bins;
        empty=empty+1;
    end
end
e_estimate=N/empty
```

## 13. randomsum.m

```matlab
%randomsum.m/created by PJNahin for "Duelling Idiots"(6/3/98)
%This m-file forms sums of terms, with each term uniformly and
%independently distributed from 0 to 1. New terms are added until
%the sum is at least equal to 1. Then, the number of terms
%required is saved and a new sum is started.
%
%

rand('state',100*sum(clock));          %new seed for generator
d=zeros(1,20);                          %initialize vector of number of
                                        %in a sum, i.e., d(i)=# of sums
                                        %that required i terms to reach
                                        %or exceed 1.
k=1;                                    %get ready to form first sum;
while k<=100000                         %have we formed last sum?
    n=0;                                %no, get ready to form new sum by
    summation=0;                        %setting both n (current number of
                                        %terms in sum), and summation to zero;

    while summation<1                   %has sum reached 1?
        summation=summation+rand;       %no, add another term;
        n=n+1;                          %up-date number of terms;
    end
    d(n)=d(n)+1;                        %yes, sum has reached threshold,
    k=k+1;                              %and so up-date number-of-terms
end                                     %vector, and the number of sums
                                        %formed so far;
e=0;                                    %start to calculated the expected
for n=1:20                              %value of the number of terms in a sum
    e=e+n*d(n);
end
e/100000                                %print simulation estimate for E(N);
```

```
%onion.m/created by PJNahin for "Duelling Idiots"(1/5/99)
%This m-file uniformly slices the unit interval, from left to right,
%ten times, and forms the sum of the widths of the slices. It does
%this 5000 times and then plots the distribution function of the sum.
%
rand('state',100*sum(clock));        %new seed for generator;
rsum=zeros(1,5000);
for cutting=1:5000
    total_thickness=0;
    L=1;
    for dice=1:10
        slice=L*rand;
        total_thickness=total_thickness+slice;
        L=slice;
    end
    index=round(100*total_thickness+.5);    %adding the '.5' prevents
    rsum(index)=rsum(index)+1;              %getting a zero index;
end
x=.01:.01:5;
dist(1)=rsum(1);
for index=2:500
    dist(index)=dist(index-1)+rsum(index);
end
plot(x,dist/5000)
grid
xlabel('thickness')
ylabel('probability')
title('Fig.5.2-Distribution of Total Thickness from 5,000 Cuttings')
figure(1)
```

## 15. match.m

```
%match.m/created by PJNahin for "Duelling Idiots"(6/3/98)
%This m-file computes the probability that N people,
%flipping fair coins n times, will each get the same
%number of heads.
%%
n=input('How many flips? ')
N=input('How many people? ')
summation=0;
for k=0:n
   bc=binomial(n,k);      %call binomial coefficient function
      summation=summation+(bc)^N;
end
summation/(2^(N*n))
```

# 16. binomial

```
function    bc=binomial(top,bottom)
%Created by PJNahin for "Duelling Idiots"(6/3/98)
%BINOMIAL computes the binomial coefficient (top)
%                                           (bottom)
%
%which is equal to: top!/((top-bottom)!(bottom)!).
%Both top and bottom must be non-negative integers
%with bottom <= top.
%
%
if top==0|top==1            %if top = 0 or 1 then
                            %top! = 1
    num=1;
else
    x=2:top;                %otherwise form 2x3x...xtop
    num=prod(x);            %to get top!
end
if bottom==0|bottom==1      %ditto for bottom!
    den1=1;
else
    x=2:bottom;
    den1=prod(x);
end
test=top-bottom;
if test==0|test==1          %ditto for (top-bottom)!
    den2=1;
else
    x=2:test;
    den2=prod(x);
```

# 17. baseball.m

```
%baseball.m/created by PJNahin for "Duelling Idiots"(10/16/98)
%This m-file calculates and plots (as a function of p) the ratio
%of the probabilities of a team winning at least 81p games out of
%81 games to the probability of winning at least 162p games out
%of 162 games, where p is the probability of winning any individual
%game.
%
res=input('Resolution of p = ')
clear x
clear ratio
for p=res:res:1-res
    if 81*p==floor(81*p)        %is 81p an integer? if yes then s81=81*p;
        s81=81*p;               %if no, use floor to round DOWN to nearest
    else                        %integer and add one (e.g., "at least 17.3
        s81=floor(81*p)+1;      %games" means "at least 18 games).
    end
    if 162*p==floor(162*p)      %ditto for s162;
        s162=162*p;
    else
        s162=floor(162*p)+1;
    end
    sumnum=0;                   %form the numerator sum;
    for k=s81:81
        bc=binomial(81,k);      %See Problem 6;
        sumnum=sumnum+bc*(p^k)*((1-p)^(81-k));
    end
    sumden=0;                   %form the denominator sum;
    for k=s162:162
        bc=binomial(162,k);
        sumden=sumden+bc*(p^k)*((1-p)^(162-k));
```

## 17. baseball.m (cont.)

```
    end
    sumden=0;                          %form the denominator sum;
    for k=s162:162
        bc=binomial(162,k);
        sumden=sumden+bc*(p^k)*((1-p)^(162-k));
    end
    index=round(p/res);
    index                              %display current value on screen
                                       %to prevent boredom!

    ratio(index)=sumnum/sumden;
    x(index)=p;
end
plot(x,ratio)
grid
title('Fig.7.1-Ratio of Probabilities of Winning At Least np Games Out of n, n=81/np=162')
xlabel('p, probability of winning any individual game')
ylabel('ratio')
figure(1)
```

## 18. balls.m

```
%balls.m/created by PJNahin for "Duelling Idiots"(12/19/98)
%This m-file calculates two quantities; (1) the average number of
%drawings (with replacement) of numbered balls before a repetition
%occurs, and (2) the largest number of drawings that still allows
%the probability there is NOT a repetition is still greater than 1/2.
% % %

n=input('How many balls are in the urn? ')
F=1/n;
E=F;
j=1;
NF=F*(n-j)/n;
while NF>0
    F=NF;
    j=j+1;
    NF=F*(n-j)/n;
    E=E+j*j*F;
end
E
j=0;
Prod=1;
while Prod > .5
    Prod=Prod*((n-j)/n);
    j=j+1;
end
T=j-1;
T
```

## 19. biased.m

```matlab
%biased.m/created by PJNahin for "Duelling Idiots"(12/27/98)
%This m-file simulates 1,000 games of odd-person-out, for N people as
%defined by the user, with N-1 persons flipping a fair coin and a single person
%person flipping a biased coin (with the probability of heads equal to
%to q). The elements of the row vector duration are the number of games of length
%i, i.e., duration(i)=# of games that require i flips to complete, where i=1,2,3...
%
%

duration=zeros(1,50);                   %initialize to zero the number of games
                                        %that have durations from 1 to 50 flips

N=input('Enter number of people playing the game:')
q=input('Enter probability of heads for the biased coin:')
rand('state',100*sum(clock))            %set new seed for the random number
                                        %generator
for game=1:1000                         %simulate 1,000 games
    gameover=0;                         %gameover=1 is the flag that the
                                        %current game is done
    flip=1;                             %initialize duration of current game
    while gameover==0                   %keep playing current game
        s=0;                            %initialize number of heads before
                                        %each flipping of N coins

        fair=N-1;                       %flip a fair coin for each of N-1
        for n=1:fair                    %people, and increment s by 1 for
            coin=rand;                  %each head
            if coin<1/2
                s=s+1;
            end
```

## 19. biased.m (cont.)

```
      end
      coin=rand;                                    %flip the biased coin;
      if coin<q                                     %increment s if biased
         s=s+1;                                     %coin shows heads;
      end
      if s==1 | s==N-1                              %is there an odd-person-out?
         gameover=1;                                %yes, set game-over flag
      else
         flip=flip+1;                               %no, so flip again
      end                                           %check while statement for end of game
      duration(flip)=duration(flip)+1; %game is over, up-date duration vector
   end                                              %play new game

average=0;
for i=1:length(duration)
   average=average+i*duration(i);
end

average=average/1000;
disp(['The average number of flips/game for ',num2str(N),' players is ',num2str(average)])
```

## 20. chess.m

```
%chess.m/created by PJNahin for "Duelling Idiots" (5/17/98)
%This m-file computes the probabilities, in an N-game chess
%match, of the match ending in a tie, in a win for the champ, or
%in a win for the challenger. The probability the champ wins an
%individual game is p, and the probability an individual
%game ends in a tie is q.
%
%
N=input('Number of games in match? ')
q=input('Probability of a drawn game? ')
ANSWER=zeros(3,round(100*(1-q)));
for p=.01:.01:1-q
    g=(1-p-q)*p;
    RZ=zeros(1,N);              %RZ plays the role of 'row zero' in the
    C=zeros(N);                 %C(j,k), as in MATLAB all matrix indices
    for k=1:N                   %start at one;
        C(k,k)=p^k;
    end
    for n=1:N
        if 2*round(n/2)===n
            for d=0:2:n
                top=(n+d)/2;
                bottom=(n-d)/2;
                I=binomial(n,bottom)*binomial(top,bottom);  %see Problem 6;
                I=I*(g^bottom)*(q^d);
                RZ(n)=RZ(n)+I;
            end
```

MATLAB Programs   231

# 20. chess.m (cont.)

```
    else
        for d=1:2:n
            top=(n+d)/2;
            bottom=(n-d)/2;
            I=binomial(n,bottom)*binomial(top,bottom);
            I=I*(g^bottom)*(q^d);
            RZ(n)=RZ(n)+I;
        end
    end
g=1-p-q;
    for c=2:N
        k=1;
        n=c;
        while n<=N
            C(k,n)=C(k,n-1)*q+C(k+1,n-1)*g;
            if k==1
                C(k,n)=C(k,n)+RZ(n-1)*p;
            else
                C(k,n)=C(k,n)+C(k-1,n-1)*p;
            end
            k=k+1;
            n=n+1;
        end
```

```
    end
SUM=0;
for k=1:N
    SUM=SUM+C(k,N);
end
Q=round(100*(1-q));
j=round(100*p);
if j<=Q
    ANSWER(1,j)=RZ(N);
    ANSWER(2,j)=SUM;
    ANSWER(3,j)=1-ANSWER(1,j)-ANSWER(2,j);
else
end
end
%
%
p=.01:.01:1-q;
plot(p,ANSWER(1,:),'h',p,ANSWER(2,:),'.',p,ANSWER(3,:),'-')
title('Fig.10.2-Tie is hexagrams, Champ wins is dots, Challenger wins is solid')
xlabel('probability, p, of the Champ winning a game')
ylabel('probability')
grid
figure(1)
```

```
%ash.m/created by PJNahin for "Duelling Idiots" (11/15/98)
%This m-file is chess.m modified to implement Ash's special
%values of p=q=1/3.
%%
%%
N=input('Number of games in match? ')
q=1/3;
ANSWER=zeros(3,round(100*(1-q)));
p=1/3;
    g=(1-p-q)*p;
    RZ=zeros(1,N);
    C=zeros(N);
    for k=1:N
        C(k,k)=p^k;
    end
    for n=1:N
    if 2*round(n/2)==n
        for d=0:2:n
            top=(n+d)/2;
            bottom=(n-d)/2;
            I=binomial(n,bottom)*binomial(top,bottom);
            I=I*(g^bottom)*(q^d);
            RZ(n)=RZ(n)+I;
        end
    else
        for d=1:2:n
            top=(n+d)/2;
```

```
      bottom=(n-d)/2;
      I=binomial(n,bottom)*binomial(top,bottom);
      I=I*(q^bottom)*(q^d);
      RZ(n)=RZ(n)+I;
   end
end
g=1-p-q;
for c=2:N
   k=1;
   n=c;
   while n<=N
      C(k,n)=C(k,n-1)*q+C(k+1,n-1)*g;
      if k==1
         C(k,n)=C(k,n)+RZ(n-1)*p;
      else
         C(k,n)=C(k,n)+C(k-1,n-1)*p;
      end
      k=k+1;
      n=n+1;
   end
end
SUM=0;
for k=1:N
   SUM=SUM+C(k,N);
end
RZ(N)
```

## 22. needle.m

```
%needle.m/created by PJNahin for "Duelling Idiots"(1/13/99)
%
%
ratio=(0:.01:1);
factor1=asin(ratio);
factor2=sqrt(1-ratio.^2);
prob1=1-(2/pi)*(ratio.*factor2+factor1);
factor3=sqrt(1-.25*ratio.^2);
factor4=asin(.5*ratio);
prob3=(2/pi)*(ratio.*factor3+2*factor4-ratio.*factor2-factor1);
prob2=1-(prob1+prob3);
figure(1)
plot(ratio,prob1)
xlabel('a/r')
ylabel('probability')
grid
title('Fig.11.3-Probability Neither End of Needle Sticks-Out Over Edge')
figure(2)
plot(ratio,prob2)
xlabel('a/r')
ylabel('probability')
grid
title('Fig.11.4-Probability One End of Needle Sticks-Out Over Edge')
figure(3)
plot(ratio,prob3)
xlabel('a/r')
ylabel('probability')
grid
title('Fig.11.5-Probability Both Ends of Needle Stick-Out Over Edge')
```

## 23. z.m

```
%z.m/created by PJNahin for "Duelling Idiots"(2/15/99)
%This m-file simulates the random variable Z=X/(X-Y),
%where X and Y are independent and uniform from 0 to 1.
%The output is a histogram of Z.
%%

rand('state',100*sum(clock));
x=-3:.05:3;
index=10000;
while index>0
    X=rand;
    Y=rand;
    Z=X/(X-Y);
    if abs(Z)<3
        vector(index)=Z;
        index=index-1;
    end
end
hist(vector,x)
title('Fig.12.5 - Histogram of Z=X/(X-Y)')
xlabel('z, in bins of width 0.05')
ylabel('number of numbers in each bin')
figure(1)
```

## 24. xpowery.m

```
%xpowery.m/created by PJNahin for "Duelling Idiots" (1/6/99)
%This m-file plots the probability density function of X^Y,
%where X and Y are independent, and both uniform from 0 to 1.
%
%
z=.0001:.01:1;
Z=expint(-log(z))./z;
plot(z,Z)
grid
title('Fig.13.2 - PDF of Z = X^Y, with X and Y uniform from 0 to 1')
ylabel('Probability Density')
xlabel('z')
figure(1)
```

## 25. xyhisto.m

```
%xyhisto.m/created by PJNahin for "Duelling Idiots"(1/6/99)
%This m-file produces 20,000 values of X raised to the Y power,
%where X and Y are both uniform from 0 to 1, and independent. It
%then plots a histogram of the values using 100 bins.
%
%
rand('state',100*sum(clock))
clear z
for pair = 1:20000
    x=rand;
    y=rand;
    z(pair)=x^y;
end
hist(z,100)
xlabel('z')
ylabel('number of values per bin')
title('Fig.13.3 - Histogram of X^Y (20,000 values)')
figure(1)
```

## 26. kids.m

```matlab
%kids.m/created by PJNahin for "Duelling Idiots"(11/12/98)
%This m-file simulates 10,000 families that have children
%until a child is born that is the same sex as the first
%child, where p is the probability of a boy.
%
rand('state',100*sum(clock))      %set new generator seed
p=input('What is the probability of a boy? ');
total=0;                          %total number of children in 10,000 families;
s=0;                              %number of children in present family;
for family=1:10000
    c=rand;                       %generate first child;
    s=s+1;
    if c<p
        first=0;                  %first=0 if first child is a boy;
    else
        first=1;                  %first=1 if first child is a girl;
    end
    while s>0
        c=rand;                   %generate another child;
        s=s+1;
        if c<p
            if first==0           %end of family growth if new
                total=total+s;    %child matches the first child
                s=0;              %in sex, i.e., is a boy;
            else
            end
```

## 26. kids.m (cont.)

```
    end
  if c>p
    if first==1                  %ditto for a girl;
      total=total+s;
      s=0;
    else
    end
  end
  end
end
total/10000                      %print average number of children in
                                 %a family;
```

## 27. tub.m

```
%tub.m/created by PJNahin for "Duelling Idiots"(2/19/99)
%This m-file computes the optimal allocation of search boats
%looking for the UNSINKABLE TUB.
%%%
ps=input('Probability a search boat finds the TUB? ');
N=input('Total number of search boats available? ');
for index=1:999
    p1=index/1000;
    p2=1-p1;
    n=(N+(log(p2/p1)/log(1-ps)))/2;
    nl=floor(n);
    nu=nl+1;
    Pl=p1*(1-(1-ps)^nl)+p2*(1-(1-ps)^(N-nl));
    Pu=p1*(1-(1-ps)^nu)+p2*(1-(1-ps)^(N-nu));
    if n>N
        n=N;
    end
    if n<0
        n=0;
    end
    if n>0&n<N
        if Pl>Pu
            n=nl;
```

## 27. tub.m (cont.)

```
    else
        n=nu;
    end
  end
  answer(index)=n;
end
p=.001:.001:.999;
plot(p,answer)
grid
ylabel('n, the number of search boats assigned to Island #1')
xlabel('p1, the probability the UNSINKABLE TUB is at Island #1')
title('Fig.16.3')
figure(1)
```

## 28. paths.m

```
%paths.m/created by PJNahin for "Duelling Idiots"(6/10/98)
%This m-file estimates the distribution and density functions
%of the random variable L denoting random path lengths
%across a unit square.
%%
rand('state',100*sum(clock))              %set new seed for generator;
length=zeros(1,142);                      %length is vector of path lengths,
for n=1:100000                            %in bins of width 0.01;
  x=rand;                                 %get a random entry point;
  angle=pi*rand;                          %get a random angle;
  A1=atan(1/(1-x));
  if angle<A1                             %pass through right vertical edge?
    L=(1-x)/cos(angle);                   %yes;
  elseif angle>A1 & angle<pi-atan(1/x)    %pass through top?
    L=1/sin(angle);                       %yes;
  else
    L=-x/cos(angle);                      %passes through left vertical edge
  end
  index=round(100*L+.5);                  %round path length to nearest .01
                                          %(adding the '.5' prevents getting
                                          %a zero index);
  length(index)=length(index)+1;          %up-date length vector;
end
x=.01:.01:1.42;
subplot(1,2,1)
```

## 28. paths.m (cont.)

```
plot(x,length/1000)
grid
xlabel('length')
ylabel('magnitude')
title('Density of L from 100,000 Paths')
subplot(1,2,2)
dist(1)=length(1);          %dist is distribution vector
for index=2:142
    dist(index)=dist(index-1)+length(index);
end
plot(x,dist/100000)
grid
xlabel('length')
ylabel('probability')
title('Distribution of L from 100,000 Paths')
figure(1)
```

## 29. theory.m

```
%theory.m/created by PJNahin for "Duelling Idiots"(6/10/98)
%This m-file plots the theoretical distribution function
%for Problem 17.
%
%
factor=2/pi;
for l = 1:1414
  lscale=l/1000;
  if l<1000
    F(l)=factor*lscale;
  else
    F(l)=factor*(1+2*acos(1/lscale)-sqrt(lscale^2-1));
  end
end
x=.001:.001:1.414;
plot(x,F)
grid
xlabel('length')
ylabel('probability')
title('Fig.17.7 - Theoretical Distribution Function')
figure(1)
```

## 30. flysquare.m

```
%flysquare.m/created by PJNahin for "Duelling Idiots"(1/19/99)
%This m-file simulates the distance-squared between 100,000
%pairs of flies, landing at random in the unit square.
%
%

rand('state',100*sum(clock))
summation=0;
for fly=1:100000
    x1=rand;
    y1=rand;
    x2=rand;
    y2=rand;
    distance2=(x1-x2)^2+(y1-y2)^2;
    if distance2>=1
        summation=summation+1;
    end
end
summation/100000
```

## 31. flycircle1.m

```
%flycircle1.m/created by PJNahin for "Duelling Idiots"(1/19/99)
%This m-file simulates the distance-squared between 100,000
%pairs of flies, landing at random in a circle with unit area.
%Random means that each fly's distance from the origin is uniform
%over a radius, and the angle of each fly's radius vector is
%uniform from 0 to 2*pi.
%
%
rand('state',100*sum(clock))
summation=0;
radius=1/sqrt(pi);
factor=2*pi;
for fly=1:100000
    r1=rand*radius;
    a1=rand*factor;
    r2=rand*radius;
    a2=rand*factor;
    x1=r1*cos(a1);
    y1=r1*sin(a1);
    x2=r2*cos(a2);
    y2=r2*sin(a2);
    distance2=(x1-x2)^2+(y1-y2)^2;
    if distance2>=1
        summation=summation+1;
    end
end
summation/100000
```

## 32. spider.m

```
%spider.m/created by PJNahin for "Dueling Idiots"(1/9/99)
%This m-file simulates 10,000 two-dimensional walks of a spider
%on a user-defined web (in the code below, the web is that
%of Figure 19.1).
%
%
rand('state',100*sum(clock))                    %new seed for generator;
t1=[3;0;3;1;2;2;3;4;2];                          %input web as COLUMN vectors;
t2=[4;0;2;1;3;5;1;1;2];
t3=[7;0;5;0;6;7;6;3;8];
t4=[8;0;8;0;0;0;8;7;0];
t5=[0;0;0;0;0;0;0;9;0];
web=[t1 t2 t3 t4 t5];                            %construct web from column vectors;
d=zeros(1,250);                                  %initialize walk duration vector;
for walks=1:10000
    position=1;                                  %put spider at starting position;
    steps=0;                                     %initialize number of steps;
    while position~=2                            %if fly not yet reached, then ...
        choices=web(position,1);                 %how many choices for next step?
        choice=floor(choices*rand)+2;            %make a random choice;
        position=web(position,choice);           %move to new position;
        steps=steps+1;                           %up-date number of steps;
    end
```

## 32. spider.m (cont.)

```
      d(steps)=d(steps)+1;      %fly reached, so up-date
end                            %duration vector;
average=0;                     %compute the average number of
for i=1:50                     %steps to reach the fly;
      average=average+i*d(i);
end
average/10000
plot(d)                        %plot duration vector;
title('Fig.19.3-Distribution of Walk Durations on the Web (10,000 walks)')
xlabel('number of steps')
ylabel('number of walks')
figure(1)
```

## 33. paradox.m

```
%paradox.m/created by PJNahin for "Duelling Idiots"(3/1/99)
%
%
clear s
clear r
a=input('What is average lifetime of single system: ')
a=1/a;
t=1:200;
for n=1:1:200
    s(n)=1-exp(-a*n);
    r(n)=s(n)*s(n)*(3-2*s(n));
end
grid
plot(t,s,'-',t,r,'.')
ylabel('probability of failure by time t')
xlabel('time t, in arbitrary units')
grid
title('Fig.20.1 - Average Lifetime of a system is 100')
legend('system','SYSTEM')
figure(1)
```

## 34. pert.m

```matlab
%pert.m/created by PJNahin for "Duelling Idiots"(11/22/98)
%This m-file executes the Critical Path Method algorithm
%modified to incorporate random task completion times for Fig.21.5.
%
rand('seed',100*sum(clock))    %new seed for generator
ntasks=8;                      %number of tasks, where task #1 is the pseudo-task
                               %BEGIN and the last task is the pseudo-task END;
                               %it is assummed that all tasks are numbered in
                               %sequence, from BEGIN as task #1 to END;
critical=zeros(1,15);          %initialize vector of critical tasks;
duration=zeros(1,20);          %initialize vector of project completion times;
data(1,:)=[0,0,0,0];           %enter individual task data, in the following
data(2,:)=[1,1,2,4,0];         %format: for task i, data(i,:)=
data(3,:)=[1,1,1,2,0];         %[<# of immediate predecessor tasks>,
data(4,:)=[2,2,3,1,2];         %<task #'s of immediate predecessors>,
data(5,:)=[2,2,3,2,4];         %<min time to do task>, <max time to do task>];
data(6,:)=[1,4,2,4,0];         %IMPORTANT POINT 1: all data statements must be
data(7,:)=[2,5,6,1,2];         %of the same length, so pad-out all statements
data(8,:)=[1,7,0,0,0];         %that have less than the maximum length with
                               %zeros.
                               %IMPORTANT POINT 2: the data statements MUST be
                               %entered in sequence from BEGIN to END, i.e.,
                               %all of the predecessor tasks of each task
                               %must appear in earlier data statements;
```

# 34. pert.m (cont.)

```matlab
%
totalruns=input('How many iterations? ');
for run=1:totalruns            %run the cpm algorithm totalruns times;
time=zeros(50,5);              %initialize matrix of important times;
p=zeros(50,50);                %initialize immediate predecessor matrix;
for i=1:ntasks                 %for task i, get number of immediate predecessors
npred=data(i,1);               %where data(i,2) to data(i,1+npred) are the task
                               %numbers of the immediate predecessors of task i;

k=1+npred;
  for j=2:k                    %tag the task pointed at by data(i,j) as
     p(i,data(i,j))=1;         %an immediate predecessor of task i, i.e.,
                               %p(i,q)=1 if task q is an immediate
                               %predecessor of task i;

  end
mintime=data(i,k+1);
maxtime=data(i,k+2);
randomtime=mintime+(maxtime-mintime)*rand;
time(i,5)=round(randomtime);   %enter time to do task i into
                               %matrix of important times

end
%Start forward pass to determine the earliest start and finish times
%for each task, stored as time(i,1) and time(i,2), respectively.
for i=1:ntasks
     earlyctime=0;             %initialize earliest possible completion
                               %time for immediate predecessors of task i
                               %(we are looking for the immediate predecessor
```

## 34. pert.m (cont.)

```matlab
                                   %with the MAXIMUM completion time, as task i
                                   %cannot be started until ALL of its predecessors
                                   %are done);

    for j=1:ntasks
        if p(i,j)==1               %if task j is an immediate
                                   %predecessor of task i then ......
            if earlyctime<time(j,2)    %if we have found a new
                                       %maximum completion time then ......
                earlyctime=time(j,2);  %up-date maximum
                                       %completion time of a
                                       %predecessor task to task i;
            else
            end
        else
        end
    end

                                   %have we checked for all
                                   %predecessor tasks?

    time(i,1)=earlyctime;              %yes, so record earliest time task i can
    time(i,2)=earlyctime+time(i,5);    %start and record earliest time task i
                                       %can be finished;
    end                                %repeat for next task;
%
%
%Start backward pass to determine the latest start and finish times
%for each task, stored as time(i,3) and time(i,4), respectively.
    projectdone=time(ntasks,1);        %get the time required to do the project
```

# 34. pert.m (cont.)

```matlab
for i=ntasks:-1:1
    lateftime=projectdone;          %initialize latest possible finish
                                    %time for task i; to do do this, look
                                    %at all of the successor tasks for
                                    %task i and find the one that starts
                                    %soonest, i.e., find the successor
                                    %task that has the MINIMUM start
                                    %time (if task i starts later than
                                    %that time then that successor task
                                    %will be delayed);

    for j=1:ntasks
        if p(j,i)==1                %if task i is an immediate predecessor
                                    %task j (i.e., if task j is an
                                    %immediate successor of task i) then ...
            if lateftime>time(j,3)  %if we have found a new minimum
                                    %start time for task j then ...
                lateftime=time(j,3);  %up-date late finish time
            else                        %for task i;
            end
        else
        end
    end                             %have we checked for all successor tasks?
    time(i,4)=lateftime;            %yes, so record latest time task i can end;
    time(i,3)=lateftime-time(i,5);  %and latest time that task i can start;
end
```

# 34. pert.m (cont.)

```
%Store results
duration(projectdone)=duration(projectdone)+1;    %up-date vector of
                                                   %project completion
                                                   %times;

realtasks=ntasks-1;
for i=2:realtasks
    slack=time(i,3)-time(i,1);
    if slack==0
        critical(i)=critical(i)+1;                 %find all tasks that have zero slack
                                                   %time and up-date critical vector;
    else
    end
end
%Display results
figure(1)
critical=critical/totalruns;                       %reduce critical to probability
bar(critical)                                      %a task is critical, and print
                                                   %bar graph of critical as
                                                   %figure 1

title('Figure 21.7 - Critical task likelihood for each task of Fig. 21.5')
xlabel('Task number')
ylabel('Probability task is critical')
figure(2)
duration=duration/totalruns;                       %reduce duration to probability;
bar(duration)                                      %print bar graph of duration as
                                                   %figure 2

title('Figure 21.8 - Range for project completion time')
xlabel('Completion time for the project')
ylabel('Likelihood')
```

## 35. generator.m

```
%generator.m/created by PJNahin for "Duelling Idiots" (1/1/99)
%This m-file plots a histogram of 100,000 random numbers
%from MATLAB's uniform generator (Version 5.2).
%
%
rand('state',100*sum(clock));        %new seed for generator;
y=rand(100000,1);                    %create column vector of 100,000
                                     %random numbers;
                                     %plot histogram of y using 100 bins;
hist(y,100)
title('Fig.A-Histogram of MATLAB 5.2 Uniform Random Number Generator')
xlabel('100 bins, 100,000 numbers')
ylabel('number of numbers in each bin')
figure(1)
```

## 36. correlation.m

```
%correlation.m/created by PJNahin for "Duelling Idiots"(1/30/99)
%This m-file plots the scatter diagram of the points (x(i),x(i+j))
%for any value of j => 0.
%
%
rand('state',100*sum(clock));    %set new seed for generator;
clear data
clear x
clear y
j=input('What is j? ');
stop=1000+j;
data=zeros(1,stop);
for i=1:stop
    data(i)=rand;
end
for i=1:1000
    x(i)=data(i);
    y(i)=data(i+j);
end
plot(x,y,'.')
xlabel('j = 9')
title('Fig.B-1000-Point Scatter Diagram for MATLAB 5.2 Uniform Random Number Generator')
figure(1)
```

37. cc.m

```matlab
%cc.m/created by PJNahin for "Duelling Idiots"(3/31/99)
%This program generates a random vector, X, of length
%20,009 and then calculates the correlation coefficient
%X(n) and X(n+j) for j = 0,1,2,...,9.
%
%
rand('state',100*sum(clock))
for i=1:20010
    X(i)=rand;
end
for j=0:9
    summation=0;
    for i=1:20000
        summation=summation+X(i)*X(i+j);
    end
    coefficient=(12*summation/20000)-3
end
```

## 38. normal.m

```
%normal.m/created by PJNahin for "Duelling Idiots"(2/27/99)
%This m-file plots a histogram of 100,000 random numbers
%from MATLAB's normal generator (Version 5.2).
%
%

rand('state',100*sum(clock));    %new seed for generator;
y=randn(100000,1);               %create column vector of 100,000
                                 %random numbers;

hist(y,100)                      %plot histogram of y using 100 bins
title('Fig.C-Histogram of MATLAB 5.2 Normal Random Number Generator')
xlabel('100 bins, 100,000 numbers')
ylabel('number of numbers in each bin')
figure(1)
```

# 39. casino.m

```matlab
%casino.m/created by PJNahin for "Duelling Idiots"(12/27/98)
%This m-file simulates 10,000 games of "Chuck-a-Luck" and
%calculates the average winnings. That is, you bet $1 each
%game and the winnings (W) are how much MORE money you have
%over and above the $10,000 betting money. The average
%winnings are, of course, W/10,000.
%
%
rand('state',100*sum(clock))           %new seed for generator;
W=0;                                    %initialize total winnings;
for games=1:10000                       %play 10,000 games;
    bet=floor(6*rand)+1;                %pick an integer from 1 to 6;
    for roll=1:3                        %roll three dice
        die1=floor(6*rand)+1;           %and determine
        die2=floor(6*rand)+1;           %what face shows
        die3=floor(6*rand)+1;           %on each;
    end
    matches=0;                          %initialize number of dice
                                        %that match your number;
    if bet==die1
        matches=matches+1;
    end
    if bet==die2
        matches=matches+1;
    end
```

# 39. casino.m (cont.)

```
    if bet==die3
        matches=matches+1;
    end
    if matches==0            %if no matches, you
        W=W-1;               %lose your $1 bet,
    else                     %but with matches you
        W=W+matches;         %win;
    end
end
W=W/10000;
disp(['Average winnings per game is ',num2str(W),' dollars'])
```

# 40. thief.m

```
%thief.m/created by PJNahin for "Duelling Idiots"(11/14/98)
%This m-file simulates "The Thief of Baghdad" problem for
%10,000 thieves.
%
%
rand('state',100*sum(clock))   %set new generator seed;
long=3;                        %time duration of long tunnel;
short=1;                       %time duration of short tunnel;
total=0;                       %total prison time for all thieves;
p1=1/3;
p2=2/3;
for thiefn=1:10000
    time=0;                    %initialize prison time for current
                               %thief;
    trytoescape=1;             %set flag for thief about to choose
                               %a door for the first time;

    while trytoescape>0
        door=rand;
        if door<p1             %thief escapes,
            total=total+time;  %so up-date total prison time;
            trytoescape=0;     %reset flag to break while loop;
        elseif door<p2         %thief picks short tunnel and so
            time=time+short;   %up-date current thief's prison time;
        else
            time=time+long;    %thief picks long tunnel;
        end
    end
end
total/10000                    %print average time in prison
```

## 41. monty.m

```
%monty.m/created by PJNahin for "Duelling Idiots" (4/8/99)
%This m-file simulates 10,000 games of "Let's Make A Deal."
%
%
rand('state',100*sum(clock))
WNS=0;                                  %initialize number of games won with NO SWITCHING;
WWS=0;                                  %initialize number of games won WITH SWITCHING;
switch1=1/3;
switch2=2/3;
for game=1:10000
    choice=rand;                        %put the prize behind one of the three doors;
    if choice <= switch1
        prize=1;
    elseif choice <= switch2
        prize=2;
    else
        prize=3;
    end
    choice=rand;
    if choice <= switch1                %pick a door;
        select=1;
    elseif choice <= switch2
        select=2;
```

## 41. monty.m (cont.)

```
    else
        select=3;
    end
    if select==prize
        WNS=WNS+1;      %if prize is behind your selection, then ...
                        %you win prize if you DON'T switch;
    else
        WWS=WWS+1;      %if prize is not behind your selection, then ...
                        %you win prize if you DO switch;
    end
WWS
WNS
```

## 42. onewaytodoit.m

```matlab
%onewaytodoit.m/created by PJNahin for "Duelling Idiots"(2/28/99)
%This m-file plots a histogram of 10,000 random numbers
%that are normally distributed with zero mean and unit variance.
%
%
rand('state',100*sum(clock));        %new seed for generator;
scale=2*pi;
for i=1:2:10000
    u=rand;
    theta=scale*rand;
    factor=sqrt(-2*log(u));
    x1=factor*cos(theta);
    x2=factor*sin(theta);
    y(i)=x1;
    y(i+1)=x2;
end
hist(y,100)                           %plot histogram of y using 100 bins;
title('Fig.D-Histogram of a Normal Random Number Generator')
xlabel('100 bins, 10,000 numbers')
ylabel('number of numbers in each bin')
figure(1)
```

# Index

expected value, 18, 31–32, 35, 56–57, 64–65, 69, 94–98, 104–6, 164, 168–69, 183–84. *See also* average

exponential integral, 51–52, 131

Fermat, Pierre de, 164
FitzGerald, George Francis, 52–53
float time. *See* slack time
floor (MATLAB command), 102, 189
FORTRAN, xvii

Gambler's Ruin problem, 164–65
gas diffusion, 37–39
Gauss, Karl Friedrich, 186
Gaussian random variable. *See* normal random variable
geometric probability, 29, 49, 53–54, 124, 129, 133–34, 155–56
geometric series, 18–19, 43–44, 85
googolhertz, 100
Gulf War, 10

harmonic series, 32, 201
Heisenberg, Werner, 14
Hertz, Heinrich, 52
histogram, 21, 162, 172–73, 182, 186–87, 196
Horowitz, Maurice, 61
hypothesis testing, 12

independence, 8–9, 20, 71, 157, 163–64, 167, 184, 186, 195
immediate predecessor tasks. *See* tasks
immediate successor tasks. *See* tasks

Jacobian determinant, 140, 195
James, F., 195
joint probability function (density or distribution), 133, 135, 138–40, 157. *See also* density function and distribution function

Kasparov, Gary, 45–46
Knuth, Donald, 194–95

Laplace, Pierre-Simon, 3, 9, 108
law of large numbers (weak), 25
Leclerc, Georges-Louis, 27
Lehmer, D. H., 178
Leibniz's rule (for differentiating an integral), 130
Lewis, W. W., 197
linear congruential generator, 178–81
Lipscombe, Trevor C., 196
lotteries, 11, 39

majority (vote-taker), 70–71
marginal probability function, 133, 139, 157. *See also* density function and distribution function
Markov, A. A., 25
Markov chain, 25, 37, 88–91
Marsaglia, George, 195
MATLAB, xii–xiv, xvii–xviii, 6, 19–20, 29, 38, 44, 52, 54, 67, 72, 77, 84–85, 87–89, 92, 94, 98–102, 111–13, 127, 131, 140, 142–43, 146–47, 160–63, 170, 179–87, 189, 192–93, 195–96
Maugham, W. Somerset, 3
maximum (of n random variables), 32
maximum likelihood, 13–14
Maxwell, James Clerk, 52
McAleer, Kevin, 22
McClellan, J. H., xviii
mean (of a random variable). *See* average
median, 168–70
Methodus Differentialis, 99
Middle-square method, 177–78
modulus, 178–80
Monte Carlo technique. *See* probabilistic simulation
Monty Hall Problem, 192–93, 197
Müller, Mervin E., 195
multiplicative generator. *See* linear congruential generator

normal random variable, 174, 185–87, 195.
null hypothesis, 13

# About the Author

Paul J. Nahin is Professor of Electrical Engineering at the University of New Hampshire/ Durham, where he teaches the undergraduate course in probability theory. During the academic year 1999–2000, he was Visiting Professor of Electrical Engineering at the University of Virginia/ Charlottesville, where he taught the graduate course in probability, random variables, and stochastic processes. He is the author of *An Imaginary Tale: The Story of* $\sqrt{-1}$ (Princeton University Press, 1998), which was awarded Honorable Mention in the Mathematics category of the 1998 Professional/Scholarly Publishing Division of the Association of American Publishers.